Basic Feedback
Controls in Biomedicine

Basic Feedback Controls in Biomedicine
Charles S. Lessard

ISBN: 978-3-031-00506-0   print

ISBN: 978-3-031-01634-9   ebook

DOI: 10.1007/978-3-031-01634-9

A Publication in the Springer series
SYNTHESIS LECTURES ON BIOMEDICAL ENGINEERING #27

Lecture #27

Series ISSN

Series Editor John D. Enderle, University of Connecticut
ISSN   1930-0328   print
ISSN   1930-0336   electronic

Basic Feedback Controls in Biomedicine

Charles S. Lessard

SYNTHESIS LECTURES ON BIOMEDICAL ENGINEERING #27

ABSTRACT

This textbook is intended for undergraduate students (juniors or seniors) in Biomedical Engineering, with the main goal of helping these students learn about classical control theory and its application in physiological systems. In addition, students should be able to apply the Laboratory Virtual Instrumentation Engineering Workbench (LabVIEW) Controls and Simulation Modules to mammalian physiology. The first four chapters review previous work on differential equations for electrical and mechanical systems. Chapters 5 through 8 present the general types and characteristics of feedback control systems and root locus, frequency response, and analysis of stability and margins. Chapters 9 through 12 cover basic LabVIEW programming, the control module with its pallets, and the simulation module with its pallets. Chapters 13 through 17 present various physiological models with several LabVIEW control analyses. These chapters cover control of the heart (heart rate, stroke volume, and cardiac output), the vestibular system and its role in governing equilibrium and perceived orientation, vestibulo-ocular reflex in stabilizing an image on the surface of the retina during head movement, mechanical control models of human gait (walking movement), and the respiratory control model. The latter chapters (Chapters 13–17) combine details from my class lecture notes in regard to the application of LabVIEW control programming by the class to produce the control virtual instruments and graphical displays (root locus, Bode plots, and Nyquist plot). This textbook was developed in cooperation with National Instruments personnel.

KEYWORDS

basic feedback controls, classical analysis, LabVIEW controls and simulation, mammalian physiological control systems

Preface

The book is intended for undergraduate students (juniors or seniors) in biomedical engineering. Most control textbooks target electrical or mechanical engineering students but do not address physiological controls. The purpose of this textbook is for the biomedical engineering students to learn some classical control theory and its application to physiological systems. In addition, the student should be able to apply the LabVIEW controls and simulation modules to mammalian physiology. The first four chapters review writing of differential equations for electrical and mechanical systems.

Chapter 5 presents the general types and characteristics of feedback control systems. Chapters 6 and 7 examine the root locus and frequency response analysis of control systems. Chapter 8 discusses the stability and margins in analysis of feedback control systems. Chapters 9–11 cover basic LabVIEW programming, the control module with its pallets, and the simulation module with its pallets, respectively. Chapter 12 presents the physiological models for control of the heart (heart rate, stroke volume, and cardiac output). In addition, the LabVIEW program is applied to analyze the cardiac ejection model and the cardiac filling model (transfer function). Chapter 13 discusses the vestibular system and its role in governing equilibrium and perceived orientation. The focus of the chapter is on the effects of the two vestibular organs: the otoliths and the semicircular canals. The two transfer functions are subsequently programmed in LabVIEW controls. Chapter 14 presents the role of the vestibulo-ocular reflex in stabilizing an image on the surface of the retina during head movement. Several computational models are presented. The human smooth pursuit transfer function is analyzed with the LabVIEW controls program. Chapter 15 presents the role of the vestibulo-ocular reflex (VOR) in stabilizing an image on the surface of the retina during head movement. Several computational models are presented. The human smooth pursuit transfer function is analyzed with the LabVIEW controls program. Chapter 16 presents the mechanical control models of human gait (walking movement); additionally, the transfer functions for the hip, knee, and ankle are analyzed with the LabVIEW controls program. The final chapter (Chapter 17) presents the respiratory control model with several LabVIEW analyses.

The text book was developed in cooperation with National Instrument personnel, in particular Eric Dean, Erik Luther, Christopher Malato, and Vivek Nath.

Contents

1. **Electrical System Equations** 1
    1.1 Kirchhoff's Laws 1
    1.2 A Review of Elements Connected in Series or in Parallel 2
    1.3 Review of Analogous Quantities 3
    1.4 Review of Source Transformations 4
    1.5 Topological Graph 5
    References 7

2. **Mechanical Translation Systems** 8
    2.1 Example 10
    2.2 Example of a Muscle Model 12
    2.3 Summary 15
    References 15

3. **Mechanical Rotational Systems** 16
    3.1 Example 18
    3.2 Gears 18
    3.3 Gear Train Example 19
    3.4 Electrical Equivalent Circuit 20
    3.5 Servomotors 21
    3.6 Armature Control Mode 21
    3.7 Field Control Mode 22
    References 24

4. **Thermal Systems and Systems Representation** 25
    4.1 Thermal Systems 25
    4.2 Mercury Thermometer Example 26
    4.3 Example of the Mercury Thermometer with Glass Added 27
    4.4 System Representation: The Block Diagram 28
    4.5 Block Diagram Example 28

4.6   Control Ratio or Transfer Function 28

4.7   The Characteristic Equation 29

4.8   Example Block Diagram of a Motor Control 30

4.9   Transfer Function Equations for the Servomotor Example 31

4.10 Signal Flow Graphs 32

4.11 Signal Flow Diagram Terminology 33

4.12 Flow Graph Algebra 33

4.13 Example of Reduction 35

References 35

5.   Characteristics and Types of Feedback Control Systems 36

5.1   Stability of a Linear System 36

5.2   Routh's Stability Criterion 36

5.3   Routhian Array 37

5.4   Example Problem: Routhian Array 38

      5.4.1 Solution 38

      5.4.2 Final solution 39

      5.4.3 Simplifying Work 40

5.5   Types of Feedback Systems 41

5.6   Static Error Coefficients 41

5.7   Steady-State Error 44

5.8   Example 46

References 48

6.   Root Locus 49

6.1   Basic Classical Methods for Analysis of Control Systems 50

6.2   Root Locus Procedures 51

6.3   Calibration of Static Loop Sensitivity 54

6.4   Rules for Construction of the Root Locus 55

6.5   Summary of Root Locus Procedures 57

6.6   Addition of Poles and Zeros 58

References 59

7.   Frequency Response Analysis 60

7.1   Steady-State Frequency Response 60

7.2   Figures of Merit Used to Measure System Performance 60

7.3   Relationship between the Root Locus and the Frequency Response ... 61

7.4   Constant Parameters on S Plane 62

7.5   Drawing the Bode Plots 66

7.6   Factors in Log Magnitude 68

7.7   Deriving the Transfer Function from the Log Magnitude 71

7.8   Summary 71

References 73

8.   **Stability and Margins** **74**

8.1   Nichols Charts 76

References 77

9.   **Introduction to LabVIEW** **78**

9.1   What Is LabVIEW? 78

9.2   Environment 78

    9.2.1   Getting Started 78

    9.2.2   Front Panel 78

    9.2.3   Block Diagram 78

    9.2.4   Controls and Indicators 79

    9.2.5   Functions/Controls Palette 81

9.3   Virtual Instruments 81

    9.3.1   Data Flow Execution 81

    9.3.2   Running a VI 83

9.4   9.4 LabVIEW Resources 84

    9.4.1   Example Finder 84

    9.4.2   Context Help 84

    9.4.3   LabVIEW Help 85

9.5   Structures/Programming Constructs 85

    9.5.1   While Loops 85

    9.5.2   For Loops 85

    9.5.3   MathScript Node 86

9.6   Data Structures 87

    9.6.1   Constants 87

    9.6.2   Arrays 87

    9.6.3   Clusters 88

9.7   Graphs and Charts 88

    9.7.1   Waveform Graph 89

    9.7.2   Waveform Chart 89

9.8   What Is the Difference? 90

9.9   Summary 91

10. Control Design in LabVIEW...92
    10.1 Control Design Functions 92
    10.2 Continuous Versus Discrete Models..92
    10.3 Model Construction 93
        10.3.1 Constructing a Transfer Function Graphically 93
        10.3.2 Constructing a Transfer Function with MathScript 93
    10.4 Model Interconnection 95
        10.4.1 Series Interconnection 95
        10.4.2 Parallel Interconnection...96
        10.4.3 Feedback Interconnection..97
    10.5 Model Analysis..98
        10.5.1 Time Response 98
        10.5.2 CD Parametric Time Analysis 98
        10.5.3 Analyzing a Step Response..99
        10.5.4 Analyzing an Impulse Response 100
        10.5.5 Frequency Response 102
    10.6 Review Exercises 103

11. Simulation in LabVIEW 105
    11.1 Simulation Loop..105
    11.2 Creating a Simulation Loop 106
    11.3 Configuring a Simulation 106
    11.4 Simulation Parameters Tab 107
    11.5 Timing Parameters Tab 108
    11.6 Generating Simulation Signals...109
    11.7 Displaying Simulation Output 110
    11.8 Implementing Transfer Functions 111

12. LabVIEW Control Design and Simulation Exercise 113
    12.1 Construction of an Open-Loop Block Diagram 115
    12.2 Construction of Closed-Loop Block Diagram........................121
    Reference..128
    LabVIEW Controls Tutorials 128

13. Cardiac Control 129
    13.1 Cardiac Parameters...129
        13.1.1 Heart Rate...129

13.1.2 Stroke Volume 130

13.1.3 Cardiac Output 131

13.1.4 Contractility 131

13.1.5 Preload and Afterload... 131

13.1.6 Autonomic Control 131

13.2 Cardiac Control Diagram.. 132

References 140

14. **Vestibular Control System** 141

14.1 Physiology and Anatomy... 141

14.1.1 Physiological Basis for Control.. 141

14.1.2 Equilibrium and Balance Control System 143

14.2 Interpretation of Block Diagram 145

14.2.1 Block Diagram of the Vestibular Control System 145

14.2.2 Block Diagram of the Semicircular Canal............................ 146

14.2.3 Block Diagram of the Otoliths.. 146

14.3 Simulation of the Control Models in LabVIEW 147

14.3.1 Transfer Function of Semicircular Canals 147

References 153

15. **Vestibulo-Ocular Control System** 154

15.1 Stimulus.. 154

15.2 Response.. 155

15.3 Normal Performance 155

15.3.1 Saccadic Eye Movements 155

15.3.2 Smooth Pursuit System 155

15.3.3 Vestibulo-Ocular Reflex and Vestibulo-Collic (Closed-Loop VCR)
       Reflexes... 156

15.4 Physiological Pathways.. 156

15.5 Special Case... 163

15.6 Computational Model... 163

15.6.1 Traditional Model: Young and Stark Model........................ 163

15.6.2 LabVIEW Computational Analysis with
       the Lisberger–Sejnowski VOR Model................................. 164

15.7 Results of the LabVIEW Analysis 168

15.8 Summary 170

References 170

16.   Gait and Stance Control System 172
    16.1   The Hip.. 175
    16.2   The Knee.. 176
    16.3   The Ankle 180
    16.4   Overall System 184
    References 185

17.   Respiratory Control System.. 186
    17.1   Pulmonary Physiology... 186
    17.2   Basics.. 186
    17.3   Method of Ventilation Control.. 186
    17.4   Gas Laws.. 188
    17.5   Gas Exchange at the Alveoli 189
    17.6   Gas Exchange in the Lungs and Tissues 189
    17.7   Gas Exchange in the Blood 190
    17.8   Conceptual Model.. 190
    17.9   Mathematical Model.. 192
    17.10 Additional Assumptions.. 195
    17.11 Derivation of Equations 195
        17.11.1 Inspiratory Muscles 195
        17.11.2 Lungs... 197
        17.11.3 Left Heart 198
        17.11.4 Brain and Tissue Transport 199
        17.11.5 Body Tissue 199
        17.11.6 Brain Tissue.. 200
        17.11.7 Body and Brain Tissue Venous Return.................................. 201
        17.11.8 Central and Peripheral Chemoreceptors 201
        17.11.9 Right Heart 202
    17.12 LabVIEW Simulations 202
    References 208

Author Biography 209

CHAPTER 1

Electrical System Equations

The purpose of this chapter is to review the different methods of writing differential equations for electrical and mechanical systems. Let use begin with a simple question, "What is a system?" A simple answer is, "A system is a combination of components that act together." However, a system may be interpreted to include physical, physiological, biological, and organizational or a combination thereof, which can be represented through common mathematical symbolism.

The methods covered in this chapter will apply only to those systems that are linear with constant parameters.

The first derivative of y with respect to time (t) is written as $dy(t)/dt$, the second derivative of y is written as $d^2y(t)/dt^2$, and the integral of y is written as follows.

$$\text{The integral of } y \equiv \int_0^t y(t)\, dt + y_0$$

where y_0 is known as the initial condition or the value of the integral at time $t = 0$; that is

$$y_0 \equiv \int_{-\infty}^0 y(t)\, dt$$

1.1 KIRCHHOFF'S LAWS

In writing equations for an electrical circuit, recall that there are two Kirchhoff's laws:

1. Kirchhoff's voltage law
2. Kirchhoff's current law

The Kirchhoff's voltage law states that the algebra of sum of the potential differences around a closed circuit must equal zero, whereas, the Kirchhoff's current law states that the summation of the currents at a junction, or node, must equal zero. Kirchhoff's voltage law tabulates the energy in a circuit, a decrease in energy is referred to as a voltage loss or drop, whereas an increase in energy is a voltage rise or gain. Kirchhoff's current law may be rephrased as: a charge that enters a node must leave that node because it cannot be stored in the node. Another way to state this law is: the currents into the node must equal currents out of the node.

For simple elements connected to a voltage source, the voltage source must equal the voltage loss in a closed circuit. For example, a resistive element connected to a voltage source results in the following equation [5]:

$$e(t) = v_R(t) = Ri(t)$$

For an inductive element, the equation is as follows:

$$e(t) = v_L(t) = L\frac{di}{dt}$$

For a capacitive element, the equation is as follows:

$$e(t) = v_c(t) = \frac{q}{C} = \frac{1}{C}\int idt = \frac{1}{C}\int i(t)$$

Connecting the resistor, inductor, and capacitor in series within a single circuit results in the following equations:

$$e(t) = v_L(t) + v_R(t) + v_c(t)$$

$$e(t) = L\frac{di(t)}{dt} + Ri(t) + \frac{1}{C}\int i(t) \text{ or}$$

$$\frac{de(t)}{dt} = L\frac{d^2 i(t)}{dt^2} + R\frac{di(t)}{dt} + \frac{i(t)}{C}$$

1.2   A REVIEW OF ELEMENTS CONNECTED IN SERIES OR IN PARALLEL

Resistors are connected in series are added as shown in the following equation:

$$R_1 + R_2 + R_3 = R_{eq}$$

For resistors connected in parallel, the conductances are added. Note that conductance is a reciprocal of resistance.

$$1/R_1 + 1/R_2 + 1/R_3 = 1/R_{eq}$$

When inductors are connected in series in the circuit, add the inductance, as shown in the following equation:

$$L_1 + L_2 + L_3 = L_{eq}$$

For inductors in parallel, add the reciprocal of the inductances:

$$1/L_1 + 1/L_2 + 1/L_3 = 1/L_{eq}$$

When capacitors are connected in series in the circuit like the resistors in parallel, add the reciprocal of conductance, as shown in the equation:

$$1/C_1 + 1/C_2 + 1/C_3 = 1/C_{eq}$$

For capacitors connected in series in a circuit, add the capacitors in the same manner as resistors in series.

$$C_1 + C_2 + C_3 = C_{eq}$$

1.3   REVIEW OF ANALOGOUS QUANTITIES

In writing either loop or notable equations, there are analogous equations corresponding to the various circuit elements. Table 1.1 presents the various analogous equations between loop (voltage) and nodal (current) equations. Recall that the resistance of an element (R) times the current (i) through the element results in the voltage (v) across the element, whereas the conductance of an element (G, the reciprocal of resistance) times the voltage (v) across the element results in current through the element [1].

TABLE 1.1: Analogous equations	
VOLTAGE	**CURRENT**
Ri	Gv
$L\dfrac{di}{dt}$	$C\dfrac{dv}{dt}$
$\dfrac{1}{C}\displaystyle\int i\,dt$	$\dfrac{1}{L}\displaystyle\int v\,dt$

1.4   REVIEW OF SOURCE TRANSFORMATIONS

Figure 1.1 illustrates the various source transformations that may be used with Kirchhoff's equations. If the electrical system equations are being written in Kirchhoff's voltage equations, then the source should be a voltage source; if it is a current source, use the source transformations to change to a voltage source [2,3].

Knowledge of writing the basic equations does not mean you know the most effective method for reducing your work. An important question facing network analyst is, "How many equations must be written to describe completely the network voltages and currents? One answer is, "As many "independent equations" as there are "unknown variables." This answer gives rise to two other questions:

1. How can the variables be selected so as to "minimize" the number of unknown variables?
2. How do you know and can assure that the equations you have written are "independent"?

The answer specific questions may appear simple; however, they are not. The answers are the following:

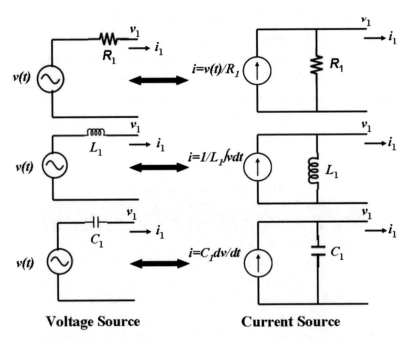

Voltage Source   **Current Source**

FIGURE 1.1: Analogous source transformations. Note that the terminals of each circuit ends with the same magnitudes of voltage and current.

1. One must write $n-1$ equations minus "known" node voltages (n nodes) where loop currents are defined as independent variables that correspond to the number of "cords."
2. A set of equation is said to be "linearly dependent" if one of the equations can be expressed as a linear combination of equation or equations.

A method often used to assist in writing the minimum number of equations is to convert the circuit network to a topological graph.

1.5   TOPOLOGICAL GRAPH

Topological graphs are networks without any circuit elements and in which straight lines replace the circuit elements. No circuit networks are "planar," meaning there are "no crossing lines." Lines of the topological graph are called "branches," and the junction of two or more branches are referred to as "nodes." A "nodal pair" is a pair of adjacent nodes. Loops are a closed path formed by connecting branches. Subgraphs that are formed after removing some of the branches for the whole graph are referred to as "trees." Necessary conditions for a tree are the following:

1. The tree must contain all the nodes of the topological graph.
2. No matter how complicated the tree is, it should contain $n-1$ branches.
3. Trees do not contain any closed paths.

Branches that are not included in the tree are referred to as "cords." The number of cords (n cords) in a topological graph is equal to n branches $-\,n$ nodes $+\,1$ [3, 4].

To construct a topological graph of a given schematic circuit, replace all circuit elements with lines, keeping and labeling only the nodes. A simple circuit diagram is shown in Figure 1.2 with its electrical elements.

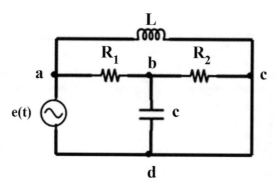

FIGURE 1.2: The circuit diagram.

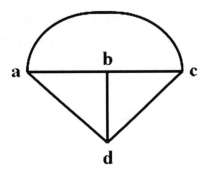

FIGURE 1.3: Topological graph. Topological graph of circuit diagram from Figure 1.2.

Figure 1.3 shows the topological graph circuit diagram from Figure 1.2. Note that all the nodes are labeled and the integrity of the diagram is maintained.

Figure 1.4 shows how the topological graph is further reduce to show the various trees and cords. The trees are drawn a solid lines, whereas the cords are shown as dashed lines [3].

In Figures 1.2–1.4, there are four nodes labeled a thorough d; thus, there are three branches, and if the Kirchhoff's nodal equations are used, $n - 1$ equations minus known node voltages must be written to solve the unknown nodal voltages. The resulting nodal equations for Figure 1.2 are written in the matrix form as follows:

$$
\begin{vmatrix} \dfrac{e(t)}{R_1} \\[2ex] 0 \end{vmatrix} = \begin{vmatrix} \left(\dfrac{1}{R_1} - \dfrac{1}{R_2} + CD \right) & -\dfrac{1}{R_2} \\[2ex] -\dfrac{1}{R_2} & \left(\dfrac{1}{R_2} - \dfrac{1}{R_2} + \dfrac{1}{LD} \right) \end{vmatrix} \times \begin{vmatrix} v_b \\[2ex] v_c \end{vmatrix}
$$

Note that there are two unknown nodal voltages v_b and v_c.

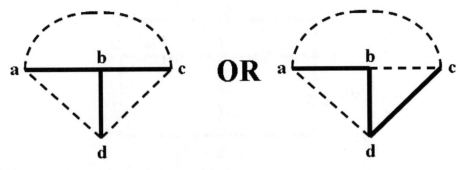

FIGURE 1.4: Various trees from the topological graph of Figure 1.3.

If Kirchhoff's loop equations are used, then the number of equations must equal the number of cords. In the case of Figure 1.2, three loop equations must be written. The first current loop includes the voltage source and encompasses nodes a, b, and d. The second current loop includes the resistor and the capacitor and encompasses nodes b–d. The third current loop includes the inductor and the two resistors and encompasses nodes a–c. The resulting loop equations for Figure 1.2 written in the matrix form are as follows:

$$\begin{vmatrix} e(t) \\ 0 \\ 0 \end{vmatrix} = \begin{vmatrix} \left(R_1 + \dfrac{1}{CD}\right) & -\dfrac{1}{CD} & -R_1 \\ -\dfrac{1}{CD} & \left(R_2 + \dfrac{1}{CD}\right) & -R_2 \\ -R_1 & -R_2 & (R_1 + R_2 + LD) \end{vmatrix} X \begin{vmatrix} i_1 \\ i_2 \\ i_3 \end{vmatrix}$$

In this particular case, it would be better to use that nodal equations and solve for the two unknown node voltages, rather than three unknown loop currents.

REFERENCES

[1] Ward, R. P., *Introduction to Electrical Engineering*, Prentice-Hall, New York, 1952.

[2] Chestnut, H., and Mayer, R. W., *Servomechanisms and Regulating Systems*, John Wiley & Sons, New York, 1954.

[3] Van Valkenburg, M. E., *Network Analysis*, 2nd ed., Prentice-Hall, Englewood Cliffs, NJ, 1964.

[4] Lewis, P. H., and Yang, C., *Basic Control Systems Engineering*, Prentice-Hall, Upper Saddle River, NJ, 1997.

[5] Phillips, C. L., and Harbor, R. D., *Feedback Control Systems*, 4th ed., Upper Saddle River, NJ, 2000.

CHAPTER 2

Mechanical Translation Systems

This chapter will review the methods of writing differential equations for translation mechanical systems. Recall Newton's law, which basically states that the sum of all forces must equal zero. Newton's law may be restated as, "The sum of applied forces must equal the sum of reactive forces." In a similar manner that resistance, inductance, and capacitance are the characterizing elements of electric systems, there are three characterizing elements in mechanical systems: mass, elastance, and damping. Similar to an electrical circuit network, a mechanical network is drawn for a mechanical system [1].

Mass (M) is the inertial element and may be written as a reactive force (f_m) equal to mass times acceleration (a), which is normally opposite the direction of the applied force. The mechanical network representation of mass is shown in Figure 2.1. One terminal labeled (a) has the motion of the mass, while the terminal labeled (b) is considered to have the motion of the reference reaction force F_m, which is a function of time and acts through the mass and is shown in the following equation in terms of frequency domain s:

$$F_m = ma = msv = ms^2x$$

where m is mass, a is acceleration, sv is the derivative of velocity, and s^2x is the second derivative of position (distance and length).

Generally, mass is always connected to reference (ground).

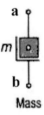

Mass

FIGURE 2.1: Mechanical representation of mass.

Spring

FIGURE 2.2: Mechanical representation of elastance element as a spring.

The elastance or stiffness element provides a restoring force as represented by a "spring" as shown in Figure 2.2. If stretched, the spring tries to contract, and if compressed, the spring tries to return to normal length. The reaction force (f_k) on each end of the spring is the same and equal to the product of the stiffness coefficient (k) of the spring and the amount of deformation of the spring [2].

Using Hook's law, the reaction or restoring force of the spring may be calculated by the following equation:

$$f_k = k(x_c - x_d)$$

Displacement of the spring is measured from the original or equilibrium position. If the end "d" is stationary, the equation simplifies to

$$f_k = kx_c$$

The final mechanical characterizing element is the damping or viscous friction (B), which is assumed to be a linear element that absorbs energy. Static friction, coulomb friction, or other nonlinear friction terms will not be included in this textbook. The damping or friction force (f_B) is proportional to the difference in the velocity of two bodies as shown in Figure 2.3. Fluid in the chambers moves from b to a. as the pressure in chamber b is greater than the pressure in chamber a due to the force F_A applied to the piston.

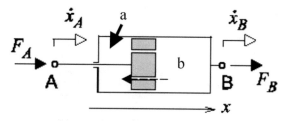

FIGURE 2.3: Mechanical representation of damping or viscous friction.

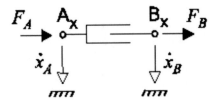

FIGURE 2.4: General representation of the mechanical damping friction element.

$$F_B = B(v_A - v_B) = B(sx_A - sx_B)$$

The reaction damping force F_B is equal to the product of the damping B and the relative velocities of the two ends of the dashpot.

The dashpot or damping friction is generally represented as shown in Figure 2.4.

2.1   EXAMPLE

Let us begin by writing the nodal equations for a simple mechanical translation system. The first step is to draw the mechanical translation system as shown in Figure 2.5. Next connect the terminals of those elements that have the same displacement (x), referred to as nodes. Figure 2.5 has three unknown nodes (x_1, x_2, and x_3) and the reference ground.

From the mechanical drawing, one can draw the mechanical network as shown in Figure 2.6. The displacements x_2 and x_3 are considered to be the same, since the center of mass for the mass element is in the middle; therefore, there are two unknown nodes and the reference node in Figure 2.6.

The next step is to write the force equations for the two unknown nodes or positions by equating the sum of the forces at each position to "zero." The equation for node x_1 is

$$f(t) = f_k = k(x_1 - x_2) = kx_1 - kx_2$$

The resulting equation for node x_2 is as follows:

$$f_k = f_m + f_B = ms^2x_2 + Bsx_2$$

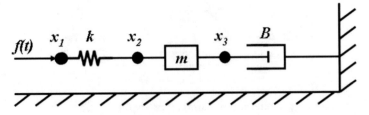

FIGURE 2.5: Simple translation system drawing.

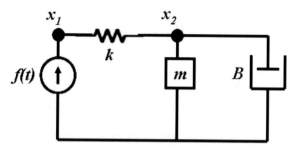

FIGURE 2.6: Mechanical network of the system shown in Figure 2.5.

The equations may be reduced to a single differential equation by substituting equation 1 for f_k in equation 2 and collecting all the x_1 and x_2 terms, which results in the following equation:

$$kx_1 = (k + ms^2 + Bs)x_2$$

In a matrix format, the two independent equations are

$$\begin{vmatrix} f(s) \\ 0 \end{vmatrix} = \begin{vmatrix} k & -k \\ -k & (ms^2 + Bs) \end{vmatrix} X \begin{vmatrix} x_1 \\ x_2 \end{vmatrix}$$

In some cases, it is desirable to simulate the mechanical system on a computer before the actual fabrication of the system. In many cases, it is easier for some students to set up an electrical analog in a laboratory. Since the mechanical nodal equations are similar to the electrical nodal equations, let us review the analogies between mechanical and electrical elements. Note that mechanical force (f) is equivalent to electrical current (i), velocity (v) is analogous to electrical voltage (v or e), and the mechanical elements are analogous to electrical component admittance: Mass (M) is equivalent to capacitance (C); the spring constant (k)is equivalent to the reciprocal of Inductance ($1/L$), and the damping or viscous friction (B) is conductance (G) or the reciprocal of resistance ($1/R$) [3].

MECHANICAL		ELECTRICAL
f		i
$v = dx/dt$		e or v
m	=	C
k		$1/L$
B		$G = 1/R$

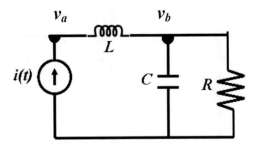

FIGURE 2.7: Analogous electrical circuit diagram.

The analogous electrical circuit diagram of the mechanical network (Figure 2.6) is shown in Figure 2.7.

The matrix equations for the analogous electrical circuit diagram are as follows:

$$
\begin{vmatrix} i(s) \\ 0 \end{vmatrix} = \begin{vmatrix} \dfrac{1}{Ls} & -\dfrac{1}{Ls} \\ -\dfrac{1}{Ls} & \left(Cs + \dfrac{1}{R}\right) \end{vmatrix} X \begin{vmatrix} v_a \\ v_b \end{vmatrix}
$$

2.2   EXAMPLE OF A MUSCLE MODEL

Physiological systems can also be modeled as mechanical or electrical networks. Let us examine the model of a muscle hanging from a beam with a mass attached to the distal end. The α-motorneuron

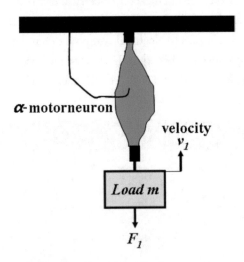

FIGURE 2.8: Model of a muscle hanging from a beam.

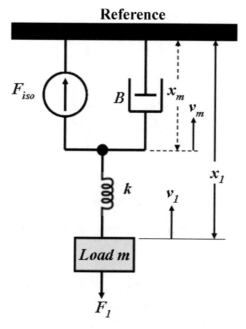

FIGURE 2.9: Equivalent mechanical network drawing of the muscle hanging from a beam as seen in Figure 2.8.

is intact and can be used to stimulate the muscle causing the muscle to contract. The mass with gravity produced a force causing the muscle to stretch. A drawing of the setup is shown in Figure 2.8, and the mechanical network is drawn in Figure 2.9. The beam is assumed to be reference since it is fixed with no velocity ($v = 0$) and all distance measurements (x) will be made referenced to the beam.

Next, the mechanical network is drawn as shown in Figure 2.10. It should be noted that the force F_{iso} is pointing toward the beam because that is the direction the muscle will contract. The mass (m) load creates a force (F_1), which stretches the muscle due to the gravitation pull of the earth. There are two unknown variables x_1 and $x_m = x_2$. The distance x_1 is measured from the mass (m) to the reference beam, while the distance ($x_m = x_2$) is measured from the node connecting the spring and the damping element to the reference beam. The electrical equivalent circuit of the mechanical network drawing is shown in Figure 2.11.

The mechanical equations for the network, written in a matrix format are as follows:

$$\left| \begin{matrix} -F_1(t) \\ -F_{iso}(t) \end{matrix} \right| = \left| \begin{matrix} \left(m\dfrac{d^2}{dt^2} + k \right) & -k \\ -k & \left(B\dfrac{d}{dt} + k \right) \end{matrix} \right| X \left| \begin{matrix} x_1 \\ x_2 \end{matrix} \right|$$

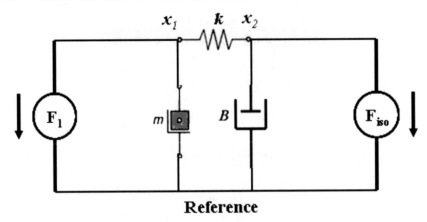

FIGURE 2.10: The mechanical network for the simple muscle system.

Note that the two sources arrows (forces $F_1(t)$ and $F_2(t)$) into the reference node are shown as negative valued, since forces into the node are labeled as positive (+) valued. Figure 2.11 is the equivalent electrical circuit of the mechanical network in Figure 2.10.

The nodal equations for the equivalent electrical circuit, written in a matrix format, are as follows:

$$\left| \begin{array}{c} -i(t) \\ -i_{iso}(t) \end{array} \right| = \left| \begin{array}{cc} \left(Cs + \dfrac{1}{sL} \right) & \dfrac{-1}{sL} \\ \dfrac{-1}{Ls} & \left(\dfrac{1}{R} + \dfrac{1}{sL} \right) \end{array} \right| X \left| \begin{array}{c} v_1 \\ v_2 \end{array} \right|$$

An alternative approach is to write the resistive element (R) as conductance (G); the result is shown in the following matrix equation for the electric circuit in Figure 2.11.

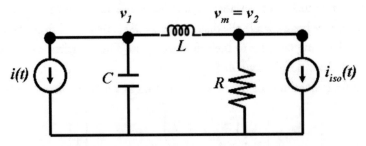

FIGURE 2.11: Equivalent electrical circuit of the mechanical network.

2.3   SUMMARY

After having reviewed the mechanical translation systems and equations, it is expected that one should be able to do the following:

1. draw the mechanical circuit network,
2. write the mechanical equations in matrix form,
3. draw the electrical analogous circuit network,
4. write the electrical analogous equations in matrix form.

REFERENCES

[1] Chestnut, H., and Mayer, R. W., *Servomechanisms and Regulating Systems*, John Wiley & Sons, New York, 1954.

[2] Lewis, P. H., and Yang, C., *Basic Control Systems Engineering*, Prentice-Hall, Upper Saddle River, NJ, 1997.

[3] Phillips, C. L., and Harbor, R. D., *Feedback Control Systems*, 4th ed., Upper Saddle River, NJ, 2000.

CHAPTER 3

Mechanical Rotational Systems

This chapter will review the methods of writing differential equations for mechanical rotational systems. Rotational systems are similar to mechanical translation systems, except that for rotational systems, "torque" equations are written rather than force equations. The linear quantities of displacement, velocity, and acceleration of mechanical translation systems are replaced with angular quantities of

1. Angular displacement (θ), in radians or degrees.
2. Angular velocity (ω), in radians/second, hertz, rpm, where angular velocity (ω) is equal to the first derivative of angular displacement (dθ). If the frequency (f) is given in hertz, then the angular velocity (ω) equals $2\pi f$.
3. Angular acceleration (α) is given in radians per seconds squared: $\alpha = $ d$^2\theta$.

Similar to the summation of forces, which must equal zero in a translation system, the "applied torque," $T(t)$, must equal the sum of the "reaction torques" in rotational systems. Rotational systems have three elements: inertial, spring, and damping elements. The reaction torque (T_J) of the inertial element (J), as shown in Figure 3.1, is often referred to as the "moment of inertia" [1–3].

The reaction torque (T_J) of the inertial element (J) has an equation similar to reaction force of mass as shown below.

$$T_J = J\alpha = J\text{d}\omega = J\text{d}^2\theta$$

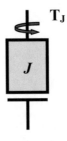

FIGURE 3.1: Inertial element (J) and its reaction torque (T_J).

FIGURE 3.2: Elastance element (k) twisted to an angular displacement (θ).

The second element is the elastance element (k), which is symbolized by a spring that is twisted to an angular displacement (Figure 3.2) rather than being stretched or compressed.

The applied torque is transmitted through the spring and appears at the other end as a reaction force. The reaction spring torque (T_k) equation is

$$T_k = k(\theta_c - \theta_d)$$

where k is the stiffness or elastance of the spring and $(\theta_c - \theta_d)$ is $\Delta\theta$ or the angle of twist of the two ends measured from the neutral position.

The third rotational element is a damping element (B), which is represented by a dashpot (Figure 3.3), as in translational system with viscous friction coefficient, B. Damping occurs when a body moves through a fluid (liquid or gas).

The damping torque (T_B) equation is shown below:

$$T_B = B(\omega_e - \omega_g) = B(d\theta_e - d\theta_g)$$

where $(\omega_e - \omega_g)$ is the relative angular velocity, $\Delta\omega$, of the ends of the dashpot.

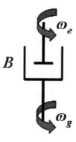

FIGURE 3.3: Rotational damping element (B) represented as a dashpot.

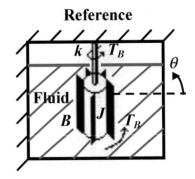

FIGURE 3.4: The simple mechanical rotational system.

3.1   EXAMPLE

Consider a simple mechanical rotational system in which a torque, T_B, is applied to a mass, J, submerged in a fluid and suspended on an elastance wire [1]. The drawing of the simple mechanical rotational system is shown in Figure 3.4.

The mechanical network drawing for the rotational system is shown in Figure 3.5.

The rotational system has one unknown variable (θ); therefore, only one torque equation is necessary. The single nodal equation is shown below:

$$T(t) = T_J + T_B + T_k$$
$$T(t) = Jd^2\theta + Bd\theta + K\theta$$

3.2   GEARS

Mechanical systems are often connected through a gear train. When a load is coupled to a gear train to a drive motor, the inertia and damping relative to the motor are important.

If the shaft is very short, the stiffness may be assumed as infinite; otherwise, elastance, k, of the shaft must be included. A symbolic drawing of a single gear train is given in Figure 3.6.

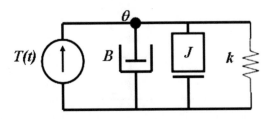

FIGURE 3.5: Mechanical network drawing for the rotational system in Figure 3.4.

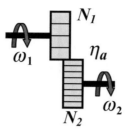

FIGURE 3.6: Symbolic drawing of a single gear train.

The mechanical advantage (η_a) of a gear train is the ratio of the driveshaft speed (ω_1) to the speed of the shaft being driven (ω_2) [2]. The mechanical management may be calculated by as:

$$\eta_a = \frac{\omega_1}{\omega_2} = \frac{\theta_1}{\theta_2} = \frac{N_2}{N_1}$$

Note that the mechanical advantage may also be calculated from the ratio of the number of teeth on each gear; however, note the inverse relationship ($\eta_a = N_2/N_1$).

3.3   GEAR TRAIN EXAMPLE

Let us examine a simple gear train example in which the input torque is applied to one end of the shaft and T_1 represents the torque load of the first gear train on the first gear train produced by the rest of the gear train. The mechanical drawing for the gear train example is shown in Figure 3.7.

The torque transmitted to the second gear train, T_2, is proportional to the product of mechanical advantage times the torque load of the first gear train, $T_2 = \eta_a T_1$. Note that the mechanical damage for these torques are inversely proportional to the speed of the gear train but are directly proportional to the number of teeth as shown in the following equations [2]:

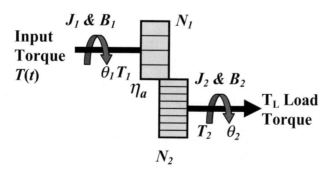

FIGURE 3.7: Mechanical drawing for the gear train example.

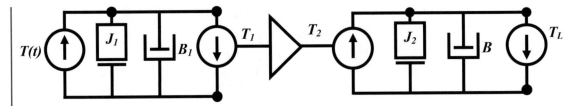

FIGURE 3.8: Mechanical network drawing of the gear train example.

$$\eta_a = \frac{T_2}{T_1} = \frac{\omega_1}{\omega_2} \quad \text{note:} \quad T_1\omega_1 = T_2\omega_2$$

$$\eta_a = \frac{T_2}{T_1} = \frac{N_2}{N_1} \quad \text{therefore,} \quad T_2 = \eta_a T_1$$

The mechanical network drawing of the gear train example is shown in Figure 3.8 [1]. The describing equations for the gear example are as follows:

$$\text{Node } \theta_1 : \; T(t) = J_1 d^2\theta_1 + B_1 d\theta_1 + T_1$$
$$\text{Node } \theta_2 : \; T_2 = J_2 d^2\theta_2 + B_2 d\theta_2 + T_{\text{load}}$$

The two equations can be reduced to a single unknown variable, because by definition

$$\theta_2 = \frac{\theta_1}{\eta_a} \quad \text{and} \quad T_2 = \eta_a T_1$$

For a gear train with two stages, $\theta_1 = \eta_a \eta_b \theta_3$.

3.4   ELECTRICAL EQUIVALENT CIRCUIT

An electrical transformer or amplifier is the electrical equivalent network of the gear train mechanical advantage as shown in Figure 3.9.

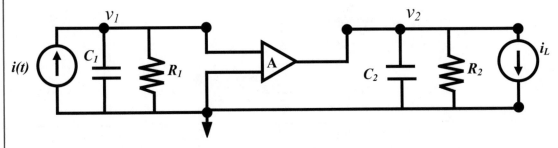

FIGURE 3.9: Electrical equivalent network of the gear train example.

If the mechanical advantage, η_a, is 1 or greater, then an amplifier (gain) is used in place of the gears; however, if the mechanical advantage is less than 1, then a potentiometer should be used in place of the gears.

3.5 SERVOMOTORS

Let us switch gears and review motors used in control systems. Recall that any current-carrying conductor located in a magnetic field experiences a force proportional to the magnitude of the magnetic flux (F), the current (i), the length of the conductor, and the sine of the angle between the conductor and the direction of the flux [1,2].

In a motor, the conductor is at a fixed distance from an axis about which the conductor can rotate. The resultant torque (T) is proportional to the product of the force (F) and the radius (r):

$$T \propto Fr$$

Because only the flux and armature current are adjustable, the resulting torque is

$$T(t) = KFi_m$$

where F is the magnetic field and i_m is the armature current.

Servomotors have two modes of operation. In one mode, the field current is held constant whereas the armature current is varied. If a constant "field current" is obtained by a separate DC excitation of the field winding, then torque is

$$T(t) = K_T i_m$$

3.6 ARMATURE CONTROL MODE

When the motor armature is rotating, a voltage is induced, which is proportional to the product of the flux and speed. The polarity of the voltage, "back emf" (e_m), is opposite to the voltage applied to the armature (e_a) [2]. Because the flux from the field (DC) is constant, the back emf is directly proportional to the armature speed (ω_m); thus,

$$e_m = K_1 F \omega_m = K_b d\theta$$

The diagram of the armature control servomotor is shown in Figure 3.10.

Control of motor speed is made possible by adjusting the applied armature voltage e_a. The polarity of e_a determines the direction of rotation. Energy losses in the armature are attributable to the armature inductance and resistance. The armature equation is

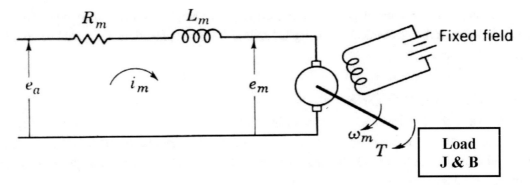

FIGURE 3.10: Armature control servomotor.

$$e_a(t) = R_m i_m + L_m(di_m/dt) + e_m$$

Current in the armature produces the required torque

$$T(t) = K_T i_m$$

For a load with only moment of inertia, J, and damper (friction), B, the equation is

$$T(t) = J d\omega_m + B\omega_m$$

By substitution, the equation becomes

$$T(t) = J d\omega_m + B\omega_m = KT i_m$$

$$i_m = \frac{J d\omega_m}{K_T} + \frac{B\omega_m}{K_T}$$

And substituting into the armature equation,

$$e_a = R_m \left(\frac{J d\omega_m}{K_T} + \frac{B\omega_m}{K_T}\right) + L_m d\left(\frac{J d\omega_m}{K_T} + \frac{B\omega_m}{K_T}\right) + e_m$$

Because the back emf is $e_m = K_b \omega_m$, the equation may be simplified to:

$$e_a = \frac{L_m J}{K_T} d^2 \omega_m + \left(\frac{R_m J + L_m B}{K_T}\right) d\omega_m + \left(\frac{R_m B}{K_T} + K_b\right) \omega_m$$

3.7 FIELD CONTROL MODE

With the second servomotor control mode, the armature current is held constant and the field current is varied. With field controls, the torque is proportional to the flux of the magnetic field, Φ, as shown in the following equation.

$$T(t) = K_3\Phi i_m = K_3\,K_2 i_f i_m$$

where $\Phi = K_2 i_f$, and the K's (K_1, K_2, K_b, etc) are proportionality constants.

With the armature current, i_m held constant, and letting $K_f = K_3 K_2 i_m$, then the simplified equation becomes

$$T(t) = K_f i_f$$

Motor speed control is obtained by varying the voltage to the "field windings" (e_f). The magnitude and polarity of e_f determine the magnitude of the torque and the direction of rotation. In this case, field inductance is usually not negligible. The electrical network for a field control servomotor is shown in Figure 3.11 [1,2].

In this example, the load has both initial J and friction B elements. The field winding equation is

$$e_f = L_f d i_f + R_f i_f$$

where the subscript "f" denotes "of the field."

The equations for the field current and field voltage are as follows:

$$i_f = \frac{J d\omega_m + B\omega_m}{K_f}$$

Then,

$$e_f = \frac{L_f D}{K_f}(J d\omega_m + B\omega_m) + \frac{R_f}{K_f}(J d\omega_m + B\omega_m)$$

or

$$e_f K_f = (L_f D + R_f)(JD + B)\omega_m$$

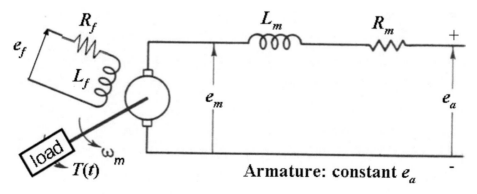

FIGURE 3.11: Field control servomotor network.

The advantage of field control is that it requires less power than armature control; however, a constant armature current i_m is difficult to achieve because of the back emf. F_f, the armature inductance, is assumed to be negligible. The full armature current equation may be reduced as follows:

$$i_m = \frac{e_a - e_m}{R_m} = \frac{e_a - K_2 \phi \omega_m}{R_m}$$

In most cases, the applied armature voltage (e_a) is much greater than e_m. Then, the armature current is simplified to the following equation:

$$i_m = \frac{e_a}{R_m}$$

The solution to electromechanical systems with capacitive coupling may be treated in several ways. Several authors use the Lagrange equations to provide a systematic approach, thus eliminating the need to consider Kirchhoff's laws and Newton's laws separately. This text will not go into a review of magnetic coupling or the development of the Lagrange equations.

REFERENCES

[1] Chestnut, H., and Mayer, R. W., *Servomechanisms and Regulating Systems*, John Wiley & Sons, Inc., New York, NY (1954).

[2] Lewis, P. H., and Yang, C., *Basic Control Systems Engineering*, Prentice-Hall, Inc., Upper Saddle River, NJ (1997).

[3] Phillips, C. L., and Harbor, R. D., *Feedback Control Systems*, 4th ed., Upper Saddle River, NJ (2000).

•   •   •   •

CHAPTER 4

Thermal Systems and Systems Representation

4.1 THERMAL SYSTEMS

Only a few thermal systems can be represented by a set of linear differential equations.

For these thermal systems, the temperature of the body is considered to be uniform, which infers perfect mixing, and that the temperature is at steady state. For equilibrium, the heat (or energy) added must be equal to the heat stored plus the heat lost.

Thermal variables are analogous to electrical elements. For example, thermal capacitance (C) is the heat storing capacity of an object, and thermal resistance (R) is the element that determines the rate of heat flow through an object in terms of two boundary temperatures. Rate of heat flow (q) is analogous to electrical current, whereas temperature (θ) is analogous to electrical voltage (potential).

Heat energy is the heat stored (h) in Joules per unit time or BTU per second as shown in the following equation.

$$h = \int q \, dt \quad \text{BTU/sec} \quad \text{or} \quad \text{J/sec}$$

where q is the rate of heat flow.

Heat energy may also be considered as the product of thermal capacitance and change in temperature, as shown in the following equation.

$$h = C(\theta_2 - \theta_1) = C(\Delta T)$$

where C is thermal capacitance (BTU/°F or J/°C) and θ or T is temperature.

The rate of heat flow (q) can be determined by using the following equations.

$$h = \int q = C(\theta_2 - \theta_1)$$

$$q = C\frac{d}{dt}(\theta_2 - \theta_1) = C\frac{d\Delta T}{dt}$$

4.2 MERCURY THERMOMETER EXAMPLE

Let us examine and characterize a thin-walled glass mercury thermometer at a stabilized room temperature $T_1 = \theta_1$; if the thin glass is assumed as negligible, the contribution of the glass elements can be neglected. The flow of heat (q) depends on the temperature in which the thermometer is exposed to, θ_i, and the temperature of the mercury, θ_m, divided by the thermal resistance of the mercury, R_m [1].

$$q_m = \frac{(\theta_i - \theta_m)}{R_m} \quad \text{analogous to} \quad i = \frac{(v_1 - v_2)}{R}$$

The thermal network drawing for the thin wall thermometer is shown in Figure 4.1. Heat into the thermometer is stored (h) in the thermal capacitance of the mercury:

$$\frac{h}{\theta_m} = \int q = C_m (\theta_m - \theta_{Ref})$$

The previous equations may be rewritten as

$$\int \frac{(\theta_i - \theta_m)}{R_m} = C(\theta_m - \theta_{Ref})$$

If $\theta_{Ref} = 0$, then

$$\theta_i - \theta_m = RC \frac{d\theta}{dt_m}$$

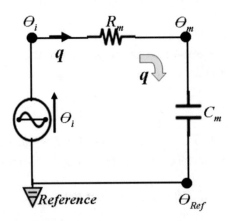

FIGURE 4.1: Thermal network drawing of a thin wall thermometer.

The thermometer differential equation may be written as a first-order system equation as follows

$$\theta_i = RC\frac{d}{dt}(\theta_m) + \theta_m$$

4.3   EXAMPLE OF THE MERCURY THERMOMETER WITH GLASS ADDED

If the glass is not thin and cannot be considered negligible, the thermal capacitance (C_g) and thermal resistance (R_g) of the glass must be added to the network diagram. In this case, the reference is assumed as equal to zero (0).

The thermal network drawing for the mercury thermometer showing the glass, thermal capacitance (C_g), and glass thermal resistance (R_g) is shown in Figure 4.2 [1].

Network equations for the glass mercury thermometer including thermal capacitance (C_g) and glass thermal resistance (R_g) in Figure 4.2 are given below in a matrix format [1].

$$\begin{vmatrix} \left(\dfrac{1}{R_g} + \dfrac{1}{R_m} + C_g\dfrac{d}{dt}\right) & -\dfrac{1}{R_m} \\[2ex] -\dfrac{1}{R_m} & \left(\dfrac{1}{R_m} + C_m\dfrac{d}{dt}\right) \end{vmatrix} \begin{vmatrix} \theta_g \\[2ex] \theta_m \end{vmatrix} = \begin{vmatrix} \dfrac{1}{R_m}\theta_i \\[2ex] 0 \end{vmatrix}$$

At this point, let us review the feedback system representation.

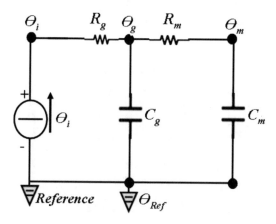

FIGURE 4.2: Thermal network drawing for a glass mercury thermometer.

4.4   SYSTEM REPRESENTATION: THE BLOCK DIAGRAM

Complete drawings of operational systems showing all system details are often too complex and congested; thus, the block diagram is used to simplify the picture of a complete closed-loop control system. Each block is labeled with component names, and the blocks are interconnected by lines. The block diagram also represents the flow of information and functions performed by each component of the system. Arrows indicate the direction of information flow. Unfortunately, the block diagram does not represent the actual physical characteristics of the system; it only shows the functional relationship between various points of the system.

4.5   BLOCK DIAGRAM EXAMPLE

An example of the block diagram is shown in Figure 4.3. In this figure, a circle used as a junction point on the diagram indicates the summation point of two or more variables of the same units. Feedback systems involve a comparison between the reference input $r(t)$ and the controlled variable $c(t)$, where the difference is the actuating signal $e(t)$ or the error signal. Differential amplifiers or potentiometers are used to measure the difference between two mechanical positions, appearing as a voltage.

4.6   CONTROL RATIO OR TRANSFER FUNCTION

The time domain representations of variables in a system block diagram are denoted with small letters as a reference to time, e.g., the reference input signal in time domain representation is $r(t)$ as in Figure 4.3. Figure 4.4 shows the block diagram representation with "frequency domain" variables, i.e., the Fourier transform of the reference input signal is denoted by capital letters in reference to the s plane.

The control ratio, often referred to as the "transfer function," is the ratio of the controlled variable, $C(s)$, to the reference input $R(s)$, where:

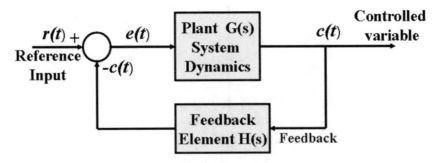

FIGURE 4.3: Typical block diagram of a feedback system.

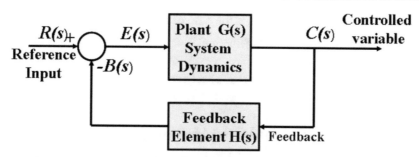

FIGURE 4.4: Frequency domain representation of a feedback system block diagram [1].

$C(s) = G(s)E(s)$ and $B(s) = H(s)C(s)$
$E(s) = R(s) - B(s)$, then $E(s) = R(s) - H(s)C(s)$
$C(s) = G(s)E(s) = G(s)R(s) - G(s)H(s)C(s)$
$C(s) + G(s)H(s)C(s) = G(s)R(s)$
$[1 + G(s) H(s)] C(s) = G(s) R(s)$

Hence, the control ratio or the transfer function is

$$C(s)/R(s) = \{G(s)/[1 + G(s)H(s)]\}$$

4.7 THE CHARACTERISTIC EQUATION

There are several definitions used in control analysis including the characteristic equation, open-loop transfer function, and forward loop transfer function. The characteristic equation of a closed-loop system is the denominator of the control ratio, i.e.,

$$1 + G(s)H(s) = 0$$

A key concept of control analysis is that the stability and response of the closed-loop system are determined by the "characteristic equation." For a simple unity feedback, $H(s) = 1$, the control ratio is:

$$\frac{C(s)}{R(s)} = \frac{G(s)}{1 + G(s)}$$

The open-loop transfer function is the ratio of the output of the feedback path, $B(s)$, to the actuating signal, $E(s)$, as shown in the following equation:

$$B(s)/E(s) = G(s)H(s)$$

The "forward loop" transfer function is the ratio of the controlled variable, $C(s)$, to the actuating signal, $E(s)$, as shown in the following equation:

$$C(s)/E(s) = G(s)$$

4.8   EXAMPLE BLOCK DIAGRAM OF A MOTOR CONTROL

In this section, the control ratio or the transfer function of a field control servomotor will be developed. Figure 4.5 shows the circuit diagram of the servomotor [1,2].

The direct current servomotor has constant field excitation and drives a frictional, inertial load.

The describing equations are shown as follows, starting with the field voltage of the generator, the generator output voltage, to the back emf of the motor. Finally, the torque is shown.

$$e_f = L_f s i_f = (Ls + R) i_f$$
$$e_g = K_g i_f$$
$$e_g - e_m = (L_g + L_m) s i_m + (R_g + R_m) i_m$$
$$e_m = K_b s \theta_o$$
$$\text{Torque} = K_T i_m = T = J s^2 \theta_o + B s \theta_o$$

The block diagram for the system is shown in Figure 4.6. The signal or data flow determines the input and output of each block, which are used to derive the transfer function of each block. The transfer functions of blocks in a series are multiplied to reduce the complexity.

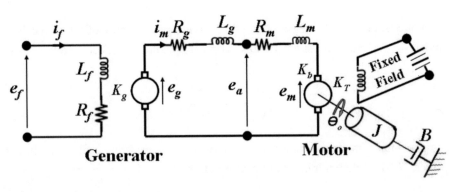

FIGURE 4.5: Circuit diagram of a field control servomotor.

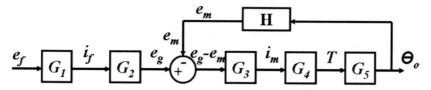

FIGURE 4.6: Block diagram of the field control servomotor.

4.9 TRANSFER FUNCTION EQUATIONS FOR THE SERVOMOTOR EXAMPLE

Let us begin with the generator field equations, and derive the control ratio (G_1) of the field current to the applied field voltage. Next, control ratios G_2, G_3, G_4, and H are derived as follows.

$$G_1 = \frac{i_f}{e_f} = \frac{1}{L_f s + R_f} + \frac{\frac{1}{R_f}}{1 + \left(\frac{Ls}{R_f}\right)s} = \frac{\frac{1}{R_f}}{1 + T_f s}$$

$$G_2 = \frac{e_g}{i_f} = K_g; \quad G_4 = \frac{T}{i_m} K_T; \quad H = \frac{e_m}{\theta_o} = K_b s$$

$$G_3 = \frac{i_m}{e_g - e_m} = \frac{\frac{1}{(R_g + R_m)}}{1 + s\left[\frac{(L_g + L_m)}{(R_g + R_m)}\right]}$$

The motor load includes the inertial (J) and viscous damping (B) elements, and must be included as part of the overall transfer function. The control ratio (G_5) of the motor inertial (J) and viscous damping (B) elements is the ratio of the output angular position, $\theta_o(s)$, to the applied torque (T) as shown in the following equation [2]:

$$G_5(s) = \frac{\theta_o(s)}{T} = \frac{\frac{1}{B}}{s\left[1 + \left(\frac{J}{B}\right)s\right]} = \frac{\frac{1}{B}}{s(1 + T_L s)}$$

The inner loop: $\quad \dfrac{\theta_o}{e_g} = \dfrac{G_3 G_4 G_5}{1 + G_3 G_4 G_5 H}$

FIGURE 4.7: Signal flow graphs showing the direction of signal flow between two nodes, x_1 and x_2, as indicated by an arrow.

In summary, the overall control ratio or transfer function G_x is as follows:

$$G_x = \frac{\theta_o}{e_f} = \frac{G_1 G_2 G_3 G_4 G_5}{1 + G_3 G_4 G_5 H}$$

4.10   SIGNAL FLOW GRAPHS

For complex systems, *signal flow graphs* are used to represent the block diagram. Signal flow graphs consist of nodes connected by directed branches and represent the system variables. Branches act as a one-way signal multiplier between two nodes. The direction of signal flow is indicated by an arrow placed on the branch, as shown in Figure 4.7. The multiplier, *a*, is indicated by a letter near the arrow and is referred to as *transmittance* [1].

The variable at node x_2 is the product of the node x_1 times the transmittance, as in the following equation.

$$x_2 = ax_1$$

Nodes perform two functions:

1. Addition of the signals from all incoming branches, and
2. Transmission of the total node signal—meaning the sum of all incoming signal to all outgoing signals, as shown in Figure 4.8.

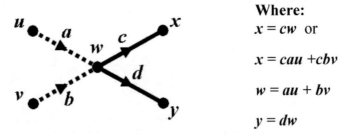

Where:

$x = cw$ or

$x = cau + cbv$

$w = au + bv$

$y = dw$

FIGURE 4.8: Example of all node incoming signals and all node outgoing signals.

FIGURE 4.9: Paths in series; nodes *xyz* are reduced into two nodes *xz* and the transmittances multiply to form *ab*.

There are three types of nodes:

1. Source nodes represent independent nodes that only have outgoing branches, i.e., nodes *u* and *v*.
2. Sink nodes represent dependent nodes that only have incoming branches, i.e., nodes *x* and *y*.
3. Mixed nodes have both incoming and outgoing branches, i.e., *w*.

4.11   SIGNAL FLOW DIAGRAM TERMINOLOGY

Specific terminology is used when describing a system *signal flow diagram*. A *path* is any connected sequence of branches whose arrows are in the same direction. A *forward path* between two nodes is one that follows the arrows of successive branches and in which nodes appears only once. In Figure 4.8, the path *uwx* is a forward path between nodes *u* and *x* [3].

4.12   FLOW GRAPH ALGEBRA

Paths in series or in parallel have different mathematical operations. *Series paths* are same as *cascade nodes*, and may be combined into a single path by multiplying the transmittances as shown in Figure 4.9.

Parallel paths, as shown in Figure 4.10, may be combined into a single path by adding the transmittances.

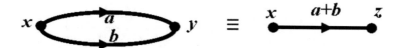

FIGURE 4.10: Paths in parallel; the transmittances add.

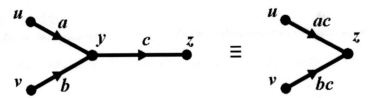

FIGURE 4.11: Node absorption; node y is eliminated by the two forward paths.

Last is the operation of *node absorption*, in which a *mixed* node may be eliminated, as shown in Figure 4.11.

To eliminate node y in Figure 4.11, the two "forward" (series) paths are reduced separately; first, the *uyz* path is reduced to *uz* with transmittance *ac*, then path *vyz* is reduced to *vz* with transmittance *bc*.

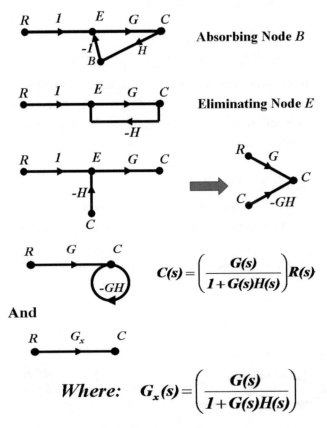

FIGURE 4.12: Sequential reduction of a feedback control system.

4.13   EXAMPLE OF REDUCTION

The following series of figures demonstrate an approach to the sequential reduction of a feedback control system to its signal flow transfer function equivalence, $G_x(s)$, as shown in Figure 4.12.

REFERENCES

[1] Phillips, C. L., and Harbor, R. D., *Feedback Control Systems*, 4th ed., Prentice-Hall, Upper Saddle River, NJ (2000).

[2] Chestnut, H., and Mayer, R.W., *Servomechanisms and Regulating Systems*, John Wiley & Sons, New York, NY (1954).

[3] Van Valkenburg, M. E., *Network Analysis*, 2nd ed., Prentice-Hall, Englewood Cliffs, NJ (1964).

CHAPTER 5

Characteristics and Types of Feedback Control Systems

So far, we have reviewed the transfer functions of open- and closed-loop systems. From the transfer function's basic characteristics, transient and steady-state analysis may be made on the feedback-controlled system.

The five most important factors in the design and performance of feedback control systems are stability, magnitude of steady-state error (error should be minimized), controllability, observability, and parameter sensitivity.

5.1   STABILITY OF A LINEAR SYSTEM

The stability of a linear system can be determined from the system's characteristic equation. If you will recall, the Heaviside expansion requires factoring of the characteristic equation, before applying a partial fraction expansion. The higher the degree of the polynomial $Q(s)$, the more laborious factoring becomes. For a stable response, all roots of the characteristic equation must lie in the left side (negative side) of the s plane.

5.2   ROUTH'S STABILITY CRITERION

Routh's stability criterion provides a simple method of determining the stability of a system without having to factor or evaluate the roots of the characteristic equation. It is not necessary to find the exact solution when the system response is unstable. In an unstable system, the characteristic equation will have one or more positive valued roots in the right side of the s plane. In addition, Routh's stability criterion may be used to determine the number of roots in the right side of the s plane with positive valued (+) real parts [1, 3].

The general form of the characteristic equation, $Q(s)$, is given as follows:

$$Q(s) = b_v s^n + b_{v-1} s^{n-1} + \cdots + b_1 s + b_0 = 0$$

where the b terms are real coefficients.

For the system to be stable, all powers of s from s_n to s_0 must be present. If any coefficient other than b_0 are zero, or if not all the coefficients are positive, then there are imaginary roots or roots with positive real parts, and the system is unstable.

If b_0 is equal to zero, divide the characteristic equation, $Q(s)$, by s.

5.3   ROUTHIAN ARRAY

To determine the number of roots in the right half of the s plane, arrange the coefficients of the characteristic equation in what is called the *Routhian array pattern*. The array pattern sets the left-most column in descending order or power of s. The first row (highest power of s) contains the coefficients (b) of every other s beginning with the highest power of s, and descending as shown below.

$$Q(s) = b_v s^v + b_{v-1} s^{v-1} + \cdots + b_1 s + b_0 = 0$$

$$
\begin{array}{c|ccccc}
s^v & b_v & b_{v-2} & b_{v-4} & b_{v-6} & \cdot \\
s^{v-1} & b_{v-1} & b_{v-3} & b_{v-5} & b_{v-7} & \cdot \\
s^{v-2} & c_1 & c_2 & c_3 & \cdot \; \cdot \; \cdot \\
s^{v-3} & d_1 & d_2 & \cdot \; \cdot \; \cdot \\
\cdot \\
\cdot \\
\cdot \\
s^1 & \dot{j}_1 \\
s^0 & k_1
\end{array}
$$

The constants (c_1, c_2, c_3, etc.) in the third row are evaluated as follows:

$$
c_1 = \frac{\begin{vmatrix} b_v & b_{v-2} \\ b_{v-1} & b_{v-3} \end{vmatrix}}{b_{v-1}} = \frac{b_{v-1} \cdot b_{v-2} - b_v \cdot b_{v-3}}{b_{v-1}}
$$

$$
c_2 = \frac{\begin{vmatrix} b_v & b_{v-4} \\ b_{v-1} & b_{v-5} \end{vmatrix}}{b_{v-1}} = \frac{b_{v-1} \cdot b_{v-4} - b_v \cdot b_{v-5}}{b_{v-1}}
$$

$$
c_3 = \frac{\begin{vmatrix} b_v & b_{v-6} \\ b_{v-1} & b_{v-7} \end{vmatrix}}{b_{v-1}} = \frac{b_{v-1} \cdot b_{v-6} - b_v \cdot b_{v-7}}{b_{v-1}}
$$

Note that in the evaluation of the two-dimensional determinate for the coefficients of c, the order is from the lower left to upper right, then from the right to lower left. This pattern is continued until the rest of the c coefficients are all equal to zero.

Next, the d coefficients are calculated as follows.

$$d_1 = \frac{c_1 b_{v-3} - b_{v-1} c_2}{c_1}$$

$$d_2 = \frac{c_1 b_{v-5} - b_{v-1} c_3}{c_1}$$

$$d_3 = \frac{c_1 b_{v-7} - b_{v-1} c_4}{c_1}$$

The rest of the rows are determined in the same manner until the values of s^1 and s^0 are calculated. The complete array forms a triangle ending with the s^0 row [1].

Once Routh's array is determined, Routh's criterion states:

"The number of roots of the characteristic equation with positive real parts is equal to the number of changes of sign of the coefficients in the first column of the array." [1]

Therefore, a system is stable if all terms in the first column of the Routh array have the same sign.

5.4   EXAMPLE PROBLEM: ROUTHIAN ARRAY

The problem requires us to:

1. calculate Routh's array
2. determine if the system is stable
3. determine the number of real parts of the roots that are in the right side of the s plane from the following characteristic equation.

$$Q(s) = s^5 + s^4 + 10s^3 + 72s^2 + 240$$

5.4.1   Solution

s^5	1	10	152
s^4	1	72	240

Start of Routh's array:

- s^3 row

$$c_1 = \frac{\begin{vmatrix} 1 & 10 \\ 1 & 72 \end{vmatrix}}{1} = \frac{1 \cdot 10 - 1 \cdot 72}{1} = \frac{10 - 72}{1} = -62$$

$$c_2 = \frac{\begin{vmatrix} 1 & 152 \\ 1 & 240 \end{vmatrix}}{1} = \frac{1 \cdot 152 - 1 \cdot 240}{1} = \frac{152 - 240}{1} = -88$$

- s^2 row

$$d_1 = \frac{\begin{vmatrix} 1 & 72 \\ -62 & -88 \end{vmatrix}}{-62} = \frac{(-62)(72) - (1)(-88)}{-62} = +70.6$$

$$d_2 = \frac{\begin{vmatrix} 1 & 240 \\ -62 & 0 \end{vmatrix}}{-62} = \frac{(-62)(240) - (1)(0)}{-62} = +240$$

- s^1 row

$$e_1 = \frac{\begin{vmatrix} -62 & -88 \\ 70.6 & 240 \end{vmatrix}}{70.6} = \frac{(70.6)(-88) - (-62)(240)}{70.6} = 122.6$$

- s^0 row

$$f_1 = \frac{\begin{vmatrix} -70.6 & 240 \\ 122.6 & 0 \end{vmatrix}}{122.6} = \frac{(122.6)(240) - (70.6)(0)}{122.6} = 240$$

5.4.2   Final Solution

Routh's array indicates that the system is unstable, because the first column of the array has two changes of sign: the first from +1 to −62 and the second from −62 to +70.6.

Routh's array

s^5	1	10	152
s^4	1	72	240
s^3	−62	−88	
s^2	70.6	240	
s^1	122.6		
s^0	240		

We can conclude that the characteristic equation, $Q(s)$, has two roots in the right side of the s plane. It should be noted that Routh's criterion does not distinguish between real and complex roots [1,3].

5.4.3 Simplifying Work

It may be occasionally necessary to simplify the work with Routh's criterion. If a coefficient in the first column of Routh's array equals zero and none of the other terms in the row are zero, then the next step is to either (1) substitute $1/x$ for s in the characteristic equation or (2) multiply the polynomial by $(s+1)$.

A useful theorem that may simplify the work states:

"The coefficients of any row may be multiplied or divided by a positive number without affecting the signs of the first column."

Let us look at another example in which the characteristic equation is given as follows:

$$Q(s) = s^4 + s^3 + 2s^2 + 2s + 5$$

The Routh array is as follows:

s^4	1	2	5
s^3	1	2	
s^2	0	5	

However, note that the determinant equation s_1 has 0 in the denominator.

$$s^1 = \frac{(2)(0) - (1)(5)}{0}$$

Substituting $1/x$ for s in the characteristic equation, the equation in terms of x becomes as follows:

$$Q(x) = 5x^4 + 2x^3 + 2x^2 + x + 1$$

Then, the array becomes

s^4	5	2	1
s^3	2	1	
s^2	$-1/2$	1	

By multiplying s^2 by +2, s^2 becomes

$$
\begin{array}{ccc}
s^2 & -1 & 2 \\
s^1 & 11 & \\
s^0 & 5 & \\
\end{array}
$$

With the characteristic equation in terms of x, $Q(x)$, the new Routh's array shows that the first column has two sign changes; therefore, the is system unstable and has two real parts of roots in the right side of the s plane.

5.5   TYPES OF FEEDBACK SYSTEMS

Designation of the type of *feedback system* is based on the order of the exponents of s. For example, the general equation for a transfer function G(s) is as follows:

$$
G(s) = \frac{K_m \left(b_w s^w + b_{w-1} s^{w-1} + \cdots + b_2 s^2 + b_1 s^1 + 1 \right)}{s^m \left(a_n s^n + a_{n-1} s^{n-1} + \cdots + a_2 s^2 + a_1 s^1 + 1 \right)}
$$

where K_m is the gain constant of $G(s)$ and m is the type of transfer function.

There are four types of feedback control systems [1–3]:

1. Type 0 system ($m = 0$; therefore, s^0) has a constant actuating signal results in a constant value for the controlled variable (constant position).
2. Type 1 system ($m = 1$; therefore, s^1) has a constant actuating signal results in a constant rate of change (constant velocity) of the controlled variable.
3. Type 2 system ($m = 2$; therefore, s^2) has a constant actuating signal results in a constant second derivative (constant acceleration) of the controlled variable.
4. Type 3 system ($m = 3$; therefore, s^3) has a constant actuating signal results in a constant rate of change of acceleration of the controlled variable.

At this point, let us turn our attention to the second most important factor in the design and performance of feedback control systems: the magnitude of *steady-state error*.

5.6   STATIC ERROR COEFFICIENTS

In their book *Servomechanisms and Regulating Systems*, Chestnut and Mayer [1] discuss about static position, velocity, and acceleration error coefficients; in contrast, D'Azzo and Houpis [4] use the terminology of static "step" instead of position, ramp instead of velocity, and parabolic instead of acceleration in defining the error coefficients in their book, *Feedback Control System Analysis and*

	TABLE 5.1: Static error coefficients		
ERROR COEFFICIENT	IDEAL TRANSFER FUNCTION	VALUE OF ERROR COEFFICIENT	FORM OF INPUT SIGNAL, $R(T)$
Position or step	K_P	$\lim_{s \to 0} (G(s))$	$R_0 u(t)$
Velocity or ramp	K_V/s	$\lim_{s \to 0} \bar{s}(G(s))$	$R_1 u(t)$
Acceleration or parabolic	K_A/s	$\lim_{s \to 0} \bar{s}^2 (G(s))$	$R_2 u(t)$

Synthesis. Definitions of the static error coefficients for a stable unity feedback system are shown as Table 5.1.

By definition, the static "step" error coefficient for a step input $r(t)$ or $R_0 u(t)$ is the ratio of the steady-state value of the output (response), $c(t)_{ss}$, to the steady-state actuating signal, $e(t)_{ss}$. Then, for type 0 system, the step error coefficient is the limit of the forward transfer function as s approaches 0; the resulting step error coefficient is K_0 as shown in the following equation.

$$\lim_{s \to 0} G(s) = \lim_{s \to 0} \frac{K_0 \prod_{n=1}^{x} (1 + T_n s)}{\prod_{m=1}^{z} (1 + T_m s)} = K_0$$

For type 0 system, note that as s approaches zero, the product terms in the numerator and denominator of the forward transfer function $G(s)$ approaches unity [1], resulting as K_0 (the final answer).

Evaluating the static "step" error coefficient for a type 1 or type 2 system results in an infinite value for the step error coefficient for either system.

The definition of the static "ramp" error coefficient for a ramp input $r(t)$ or $R_1 t u(t)$ is the ratio of the steady-state value of the derivative of the output (response), $d/dt(c)_{ss}$ to the steady-state actuating signal, $e(t)_{ss}$. Then, for type 0 system, the ramp error coefficient is the limit of the derivative of the forward transfer function as s approaches 0, thus resulting in the ramp error coefficient equal to zero (0) as shown in the following equation.

$$\lim_{s \to 0} sG(s) = \lim_{s \to 0} \frac{sK_0 \prod_{n=1}^{x} (1 + T_n s)}{\prod_{m=1}^{z} (1 + T_m s)} = 0$$

For type 0 system, note that as s approaches zero the numerator goes to zero (0), whereas denominator of the terms approaches unity (1), resulting in zero (0) as the final answer. For a type 1 system, the ramp error coefficient is a constant equal to K_1; for a type 2 system, the ramp error coefficient approaches infinity.

The definition of the static "parabolic" error coefficient for a parabolic input $r(t)$ or $R_2 t^2 u(t)$ is the ratio of the steady-state value of the second derivative of the output (response), $d^2/dt^2(c)_{ss}$ to the steady-state actuating signal, $e(t)_{ss}$. Then, for type 0 system, the parabolic error coefficient is the limit of the second derivative of the forward transfer function as s approaches 0, thus resulting in the parabolic error coefficient equal to zero as shown in the following equation.

$$\lim_{s \to 0} s^2 G(s) = \lim_{s \to 0} \frac{s^2 K_0 \prod_{n=1}^{x} (1 + T_n s)}{\prod_{m=1}^{z} (1 + T_m s)} = 0$$

For type 0 and type 1 systems, it should be noted that as s approaches zero the numerator goes to zero (0), whereas the denominator of the terms goes to unity (1), resulting in the value of the parabolic error coefficient of zero as the final answer for both systems. For a type 2 system, the ramp error coefficient is a constant equal to K_2.

The static error coefficients are summarized in Table 5.2 [1].

TABLE 5.2: Summary of static error coefficient results			
SYSTEM TYPE	STEP ERROR COEFFICIENT	RAMP ERROR COEFFICIENTS	PARABOLIC ERROR COEFFICIENT
0	K_0	0	0
1	∞	K_1	0
2	∞	∞	K_2

5.7   STEADY-STATE ERROR

Design and test engineers are usually interested in this basic question: "What type of steady-state error would result in a particular type of control system for a particular input?" The answer lies in the analysis of two theorems:

1. Final value theorem
2. Differentiation theorem

The following equation is used to analyze the final value theorem.

$$\lim_{t \to \infty} f(t) = \lim_{t \to 0} sF(s)$$

With all initial conditions set to zero, the equation for analysis of the differentiation theorem is as follows.

$$\varsigma \frac{[d^m c(t)]}{dt^m} = s^m C(s)$$

Let us consider an example to determine the steady-state output for a unity feedback closed-loop system, where the forward transfer function defines the type of system. The general block diagram for the example is shown in Figure 5.1.

The general equation for the forward loop transfer function of the unity feedback closed-loop system in Figure 5.1 is as follows:

$$G(s) = K_m \frac{(1 + T_1 s)(1 + T_2 s)(1 + T_3 s)(\ldots)}{s^m (1 + T_a s)(1 + T_b s)(\ldots)}$$

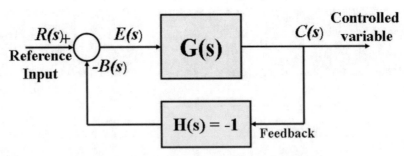

FIGURE 5.1: General block diagram for a unity feedback closed-loop system.

The actuating signal, input to the forward loop, is $E(s)$ has the following form.

$$G(s) = \frac{C(s)}{E(s)} \quad \text{or} \quad E(s) = \frac{C(s)}{G(s)}$$

Then the steady-state error is the limit of the derivative of the actuating or error signal.

$$e(t)_{ss} = \lim_{s \to 0} [sE(s)]$$

It should be noted that the derivative of the output is related to the steady-state error. Rewriting the steady-state error equation in terms of the forward transfer function results in

$$e(t)_{ss} = \lim_{s \to 0} s \left[\frac{\prod\limits_{n=1}^{x} (1 + T_n s)}{\prod\limits_{m=1}^{z} (1 + T_m s)} \right]$$

When the final value theorem is applied, the following result is obtained.

$$e(t)_{ss} = \lim_{s \to 0} \frac{s\,[s^m C(s)]}{K_m} = \frac{d^m c(t)_{ss}}{K_m}$$

Rewriting the last equation by multiplying the equation with the constant K_m results in the following:

$$K_m e(t)_{ss} = \frac{d^m}{dt^m} c(t)_{ss}$$

When $d^m/dt^m c(t)_{ss}$ is constant, then $e(t)_{ss}$ must also be constant. Engineers often prefer to relate the steady-state error to the input reference signal, $R(s)$; hence, the equations are as follows.

$$C = \left[\frac{G}{1 + G} \right] R \text{ for } E = \frac{C}{G}$$

$$E = \left(\frac{1}{G} \right) \left(\frac{G}{1 + G} \right) R = \left[\frac{R}{1 + G} \right]$$

For unity feedback:

$$C(s) = \left[\frac{\dfrac{K_m \prod (1 + T_z s)}{s_m \prod (1 + T_p s)}}{1 + \dfrac{K_m \prod (1 + T_z s)}{s_m \prod (1 + T_p s)}} \right] R(s)$$

Canceling denominators:

$$C(s) = \left[\frac{K_m \prod(1 + T_z s)}{s_m \prod(1 + T_p s) + K_m \prod(1 + T_z s)} \right] R(s)$$

Then,

$$E = \left(\frac{1}{G} \right) \left(\frac{G}{1 + G} \right) R$$

For unity feedback:

- Canceling terms

$$E(s) = \left[\frac{s_m \prod(1 + T_p s)}{K_m \prod(1 + T_z s)} \right] \left[\frac{K_m \prod(1 + T_z s)}{s_m \prod(1 + T_p s) + K_m \prod(1 + T_z s)} \right] R(s)$$

$$e(t)_{ss} = \lim_{s \to 0} \left[\frac{s_m \prod(1 + T_p s)}{s_m \prod(1 + T_p s) + K_m \prod(1 + T_z s)} \right] R(s)$$

The steady-state error is related to the input.

5.8   EXAMPLE

Let us determine the steady-state error for a type 0 control system with a step input signal.

Type 0 system: $m = 0$ and $s^0 = 1$

Step input $\mu(t) = \dfrac{R_0}{s}$

$$e(t)_{ss} = \lim_{s \to 0} \frac{sR_0}{s} \left[\frac{\prod(1 + T_p s)}{\prod(1 + T_z s) + K_0 \prod(1 + T_p s)} \right]$$

Cancel s/s

All $\prod(1 + T_z s)$ and all $\prod(1 + T_p s) \to 1$

Then,

$$e(t)_{ss} = \frac{R_0}{1 + K_0} = \text{Constant} \quad E_0 \neq 0$$

Note that for a type 0 control system with a step input signal, the steady-state error is a constant. Let us change the input to the type 0 control system into a ramp input signal.

$$R_1 t\mu(t) = \frac{R_1}{s^2}$$

$$e(t)_{ss} = \lim_{s \to 0} \frac{sR_1}{s^2} \left[\frac{1}{1 + K_1} \right]$$

$$e(t)_{ss} = \lim_{s \to 0} \frac{R_1}{s(1 + K_1)} = \infty$$

Note that in the last equation, the steady-state error $e(t)_{ss}$ has an s in the denominator that in the limit goes to zero; hence, the steady-state error for a ramp input in the limit goes to infinity. What this means is that a type 0 system cannot follow a ramp input.

The same is true for a type 0 control system with an acceleration input, $R_2 t^2 \mu(t) = R_2/s^3$, as shown in the following equation. Again, note that the following equation has an s in the denominator that in the limit goes to zero; hence, the steady-state error for a ramp input in the limit goes to infinity, which means that a type 0 system cannot follow an acceleration input.

$$e(t)_{ss} = \lim_{s \to 0} \frac{R_2}{s^2(1 + K_2)} = \infty$$

Table 5.3 presents the resulting steady-state error for the three types of control systems and the three types of input signals.

From Table 5.3, one may conclude that a type 1 system:

1. Follows a position or step input signal with zero error;
2. Has a constant error for a velocity or ramp input;
3. Cannot follow an acceleration or parabolic input.

TABLE 5.3: Summary of steady-state error $[e(t)_{ss}]$ results			
SYSTEM TYPE	POSITION OR STEP INPUT	VELOCITY OR RAMP INPUT	ACCELERATION OR PARABOLIC INPUT
0	$K_0 R_0/(1 + K_0)$	∞	∞
1	0	R_1/K_1	∞
2	0	0	R_2/K_2

Likewise, from Table 3, it is concluded that a Type 2 system:

1.   Follows a step input signal or a ramp input signal with zero (0) error;
2.   Has a constant error for an acceleration input signal.

REFERENCES

[1]   Chestnut, H., and Mayer, R. W., *Servomechanisms and Regulating Systems*, John Wiley & Sons, Inc., New York, NY (1954).

[2]   Lewis, P. H., and Yang, C., *Basic Control Systems Engineering*, Prentice-Hall, Inc., Upper Saddle River, NJ (1997).

[3]   Phillips, C. L., and Harbor, R. D., *Feedback Control Systems*, 4th ed., Upper Saddle River, NJ (2000).

[4]   D'Azzo, J. J., and Houpis, C. H., *Feedback Control System Analysis and Synthesis*, 2nd ed., McGraw-Hill Book Company, New York (1966).

•   •   •   •

CHAPTER 6

Root Locus

In this chapter, the basic concept and creation of the root locus are presented. Even though better computer analysis methods have been developed, it is the author's goal to go beyond using a software program that the user never knew about, understood, or cared about. Therefore, the basic rules that were used in the manual development of the root locus will be presented as well as the use of a Laboratory Virtual Instrumentation Engineering Workbench (LabVIEW) control program for analysis with the root locus.

Most engineers can design and put together a control system; however, few can determine if the control systems actually meet specifications. In most cases, it is necessary to know the desired time response of the controlled variable, which can be obtained by deriving the differential equations for the control system, and solving the differential equations to obtain an accurate solution for the system's performance.

For simple systems that meet specifications, the task may be simple; however, if the system does not meet specifications, then it may be difficult to determine the solution. The difficulty may be in determining *which* or *what* physical parameters of the system should be changed to improve the response in order to meet the specifications. Design engineers would prefer to predict the system's performance by any analysis method that does not require the actual solution of the differential equation. The analysis should indicate the manner in which the system must be adjusted or compensated to produce the desired performance characteristics.

The following question arises: "What should be known about a control system's time response?"

The performance of a system may be evaluated in terms of the following quantities [1]:

1. Maximum overshoot, c_p, is the magnitude of the first overshoot, which may also be expressed in percent of the final value.
2. Time to maximum overshoot, t_p, is the time required to reach the maximum overshoot.
3. Time to first zero error, t_0, is the time required to reach the final value the first time. It is often referred to as duplicating time.
4. Settling time, t_s, is the time required for the output response first to reach and thereafter

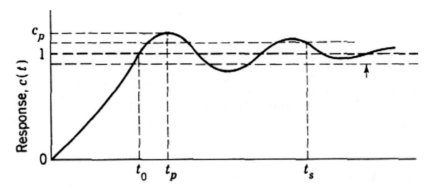

FIGURE 6.1: Typical underdamped response to a step function.

remain within a prescribed percentage of the final value. Common values used for settling time are 2% and 5% and is applied to the envelope that yields t_s.

5.   Frequency of oscillation of the transient, ω_d.

   The time response will differ for each set of initial conditions; therefore, to compare the time response of various systems, it is necessary to start with the same standard initial conditions. The most practical standard is to start with the system at rest. A typical underdamped response to a step input into a second-order system will oscillate before settling down to some constant value, as shown in Figure 6.1. The arrow at the right side of the figure points to the allowable tolerance (horizontal dashed line) set in the specifications.

   Let us not forget that it is essential to first determine the stability of a control system. As discussed in the previous chapter, the stability of a system can be determined by applying Routh's criterion to the characteristic equation; however, Routh's criterion cannot determine the following:

1.   Degree of stability
2.   Amount of overshoot
3.   Settling time of the controlled variable

6.1   BASIC CLASSICAL METHODS FOR ANALYSIS OF CONTROL SYSTEMS

There are two basic methods for analysis and interpretation of the steady-state sinusoidal response of the control system's transfer function. The first method is based on the interpretation of the system's Nyquist plot. The second method is the root locus approach, which incorporates the more desirable features of both the classical solution to the differential equation and the frequency response method.

The root locus is a plot of the characteristic equation (roots) of the closed-loop system as a function of gain. The root locus is a graphical approach that yields a clear indication of the effects of gain adjustment with the least amount of effort as compared with other methods. The underlying principle is based on the fact that the poles of the control ratio, $C(s)/R(s)$, are related to "zeros" and "poles" of the open-loop transfer function, $G(s)H(s)$, and the gain of the system.

The advantage of the root locus method is that it results in a complete solution of the controlled variable for by yielding both the "transient response" and the "steady-state response." In addition, the root locus can be used to synthesize a compensator, which is easy if a computer with the LabVIEW "Control and Simulation" software module is available.

6.2   ROOT LOCUS PROCEDURES

The major steps that engineers have used in developing the root locus are as follows [2]:

1. The system describing equations in the time domain are derived and the block for the feedback system is drawn. Because it is often easier to work in the frequency domain, the Laplace transform is taken of the describing equations and the Fourier transform is taken of the input signal.
2. Derive the open-loop transfer function $G(s)H(s)$ of the feedback system.
3. Factor the numerator and denominator into linear factors of the form $(s + a)$.
4. Plot the zeros (factors of the numerator) and poles (factors of the denominator) of the open-loop transfer function in the $s = \sigma + j\omega$ plane. The real term (σ) of the s plane is expressed in units of "neper frequency," whereas the imaginary term $(j\omega)$ is in units of "radian frequency."
5. The plotted zeros and poles of the open-loop transfer function determine the roots of the characteristic equation of the closed-loop system $[1 + G(s)H(s) = 0]$. In the days before computers, a "spirule" was used to obtain points on the root locus. At present, desktop computers with a control program software (e.g., LabVIEW) can draw the transfer function root locus.
6. Calibrate the locus in terms of the static loop sensitivity, K (with the coefficients of s all equal to unity). If the gain is predetermined, then the final location of the exact roots of $[1 + G(s)H(s)]$ is known.
7. Once the roots have been found the equation for the system's time response can be calculated by taking the inverse Laplace transform.

8. If the response does not meet the specifications, determine the shape that the root locus must have in order to meet the desired specifications.
9. Compensate the system if other then gain is required.

One should keep in mind that the underlying principle of the root locus method is based on the fact that the poles of the control ratio, $C(s)/R(s)$, are related to the zeros and poles of the open-loop transfer function $G(s)H(s)$ one and to the static loop sensitivity.

For the static loop sensitivity greater than zero, $K > 0$, the open-loop transfer function has the form:

$$\frac{C(s)}{R(s)} = \frac{G}{1 + GH}$$

where the roots are

$$\beta(s) = 1 + G(s)H(s) = 0$$

Then

$$G(s)H(s) = -1$$

Or

$$G(s)H(s) = \frac{K(s - Z_1)\dots(s - Z_w)}{s^n(s - P_1)\dots(s - P_x)} = -1$$

where K is the static loop sensitivity for $K > 0$.

The open-loop transfer function as the exponential form is as follows:

$$G(s)H(s) = -1 = Fe^{-j\beta}$$

And

$$-1 = |1| e^{j(1 + 2m)\pi} \text{ for } m = \pm 1, \pm 2, \pm \dots$$

Then, $G(s)H(s)$ is rewritten as:

$$G(s)H(s) = -1 = Fe^{-j\beta} = |1| e^{j(1 + 2m)\pi}$$

where
F, the magnitude condition of $G(s)H(s)$ must equal unity [2]
β, the angle condition of $G(s)H(s)$, must equal π (or 180°) or an odd multiple of π or 180°

The two conditions F and β determine the points that lie on the root locus and can be summarized as follows:

1. Magnitude condition for $K > 0$ open-loop $G(s)H(s) = 1$
    "The magnitude of the open-loop transfer function must always be *unity*."
2. Angle condition for $K > 0$
    $G(s)H(s) = (1 + 2n)180°$ for $n = 0, 1, 2, \ldots$
    "The phase angle of the open-loop transfer function must be an odd multiple of 180°."

Example

Let us examine an example in which the problem is in determining the locus of all possible closed-loop poles for the following open-loop transfer function.

$$G(s)H(s) = \frac{K_0(1 + 0.25s)}{(1 + s)(1 + 0.5s)(1 + 0.2s)}$$

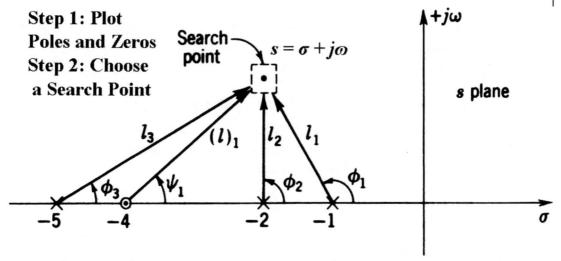

$$\angle -\beta = \sum\angle \text{(numerator terms)} - \sum\angle\text{(Denominator Terms)}$$
$$\angle +\beta = \sum\angle\text{(Pole Terms)} - \sum\angle\text{(Zero terms)}$$
$$\angle +\beta = \sum (\Phi_1 + \Phi_2 + \Phi_3) - \Psi_1$$

FIGURE 6.2: Locating a search point that lies on the root locus.

Rewriting $G(s)H(s)$ with the coefficients of s all equal to unity, results in the following equation.

$$G(s)H(s) = \left[\frac{(0.25)K_0}{(0.5)(0.2)}\right]\left[\frac{(s+4)}{(s+1)(s+2)(s+5)}\right]$$

$$= \frac{2.5K_0(s+4)}{(s+1)(s+2)(s+5)}$$

Setting the static loop sensitivity, $K = 2.5K_0$, the equation becomes:

$$G(s)H(s) = \frac{K(s+4)}{(s+1)(s+2)(s+5)}$$

After plotting poles and zeros in the s plane, and after selecting (guessing) the location of a search point, draw the directed line segments from all open-loop poles and zeros to search point as shown in Figure 6.2.

Next, check to make sure that the search point satisfies the angle condition.

$$\beta = \sum(\phi_1 + \phi_2 + \phi_3) - \psi_1 = \left\{\begin{array}{l}(1+2m)\,180°\,;\, K > 0 \\ m(360°)\,;\, K < 0\end{array}\right\}$$

If the angle condition is not satisfied, select another search point until the angle condition is satisfied. By trial and error, locate a sufficient number of points in the s plane that satisfy the angle condition (β) to completely draw the root locus.

6.3   CALIBRATION OF STATIC LOOP SENSITIVITY

The next step is to calibrate the root locus in terms of static loop sensitivity. From a search point that satisfied the angle condition, calculate the gain from the known directed line segments to the search point (Figure 6.2) using the following equation

$$|K| = \frac{l_{p_1} \cdot l_{p_2} \cdot l_{p_3}}{l_{z_1}}$$

where $l_{p_1} = |s+1|$, etc., and $l_{p_1} = |s+4|$.

Because the complex roots occur in conjugate pairs and the locus is symmetrical about the real axis, the bottom half of the locus can be drawn once the locus above the real axis has been determined. As noted earlier, there are programs (e.g., LabVIEW Control and Simulation Module, National Instruments) that obtain the root locus plot, frequency response plots, transient response plot, etc.

6.4   RULES FOR CONSTRUCTION OF THE ROOT LOCUS

Let us examine the rules for manual construction of the root locus [2].

Rule 1: *Number of branches*. The number of branches of the root locus equals the number of poles of the open-loop transfer function, $G(s)H(s)$, or the degree of the characteristic equation of the closed-loop system.

Rule 2: *Real-axis locus*. Apply the angle condition to a search point on the real axis. It should be noted that:

1.   All poles and zeros to the "left" of the search point contribute 0°.
2.   The angular contribution of complex conjugate poles to the search point is 360°.
3.   All poles and zeros to the "right" of the search point contribute 180°.
4.   For the search point on the real axis to satisfy the angle condition, the total number of poles and zeros to the "right" of search point is must be odd.

Rule 3: *Locus end points*. When $s = p_n$ the numerator becomes zero and the static loop sensitivity (K) at open-loop poles is zero, and when $s = z_m$ the denominator becomes zero and the static loop sensitivity (K) at open-loop zeros is infinity.

Rule 4: *Asymptotes of locus as* s *approaches infinity*. The following equation means that regardless of what magnitude s may have after it has reached a sufficiently large value, the angle $\angle s = \gamma°$ remains constant; however, the asymptote angle does not tell where the asymptote intercepts the real axis.

$$\gamma° = \frac{(1 + 2m)180°}{(\text{\# of poles } - \text{ \# of zeros})} \text{ as } s \to \infty$$

Rule 5: *Real-axis intercept of asymptotes*. The real-axis asymptote intercept is based on the difference of the summation of real parts of zeros subtracted from the summation of the real parts of poles divided by the difference of the total number of zeros (w) subtracted from the total number of poles (v) in the open-loop transfer function as follows [2,3]

$$z_o = \sigma_o = \frac{\sum_{c=1}^{v} \text{Real part of poles } (p_c) - \sum_{m=1}^{\omega} \text{Real part of zeros } (z_m)}{v - \omega}$$

where $z_o = \sigma_o$ is the centroid of the pole zero on the real axis.

Rule 6: *Breakaway point on real axis*. At the poles, the starting value of the static loop sensitivity is $K = 0$, and increases as it moves away from the pole. Somewhere in between, K's for the two branches simultaneously reach a maximum value. To determine the breakaway and/or break-in points on the real axis, let $W(s) =$ the denominator of the transfer function of the characteristic equation, $G(s)H(s) = -K$, then take the derivative $d[W(s)]/ds$ to determine maximums and minimums of the derivative. The peaks or maximums identify breakaway points, and the troughs or minimums identify break-in points, K's.

Let us look at an example.

$$G(s)H(s) = \frac{K}{s(s + 1)(s + 2)} = -1$$

$$W(s) = s(s - 1)(s + 2) = -K$$

$$W(s) = s^3 + 3s^2 + 2s = -K$$

$$\frac{dW(s)}{d(t)} = 3s^2 + 6s + 2 = 0$$

$$s_{1,2} = -1 \pm 0.57$$

$$s_1 = -0.43 \text{ and } s_2 = -1.57$$

The results for the derivation, for $K > 0$, indicate that the breakaway point is between $s = 0$ and $s = 1$ at $s_1 = -0.43$, and that the other point at s_2 is the break-in point at -1.57 for $K < 0$.

Rule 7: *Complex poles—angle of departure.* To determine the angle of departure from complex poles, apply the angle condition at one of the complex poles. The rule states that the direction of the locus as it leaves a pole or zero can be determined by adding all angles of the vectors from all the other poles and zeros to the pole or zero in question, then subtracting the resulting sum from 180°. For example, the point of interest, p_2, with angle of departure, φ_2. Applying the angle conditions results in the following.

$$\phi_0 + \phi_1 + \phi_2 + \phi_3 - \psi_1 = (1 + 2m)180^{\circ}$$

$$\phi_2 = (1 + 2m)180^{\circ} - (\phi_0 + \phi_1 + \phi_3 - \psi_1)$$

where Φ_2 is the angle the locus will follow after leaving p_2.

Rule 8: *Imaginary-axis crossing point.* The crossover point is where the locus crosses the imaginary axis into the right half of the s plane. The imaginary axis crossing point is determined by Routh's array method. If the s_1 row of the Routh's array is equal to zero, then an undamped oscillation exists, which means the system may cross the y axis. At this point, the auxiliary equation must be used for row s_2 of the Routhian array.

Example

Determine the imaginary axis crossing point for the general closed-loop characteristic equation given as follows:

$$as^3 + bs^2 + cs + Kd = 0$$

where $a = 1$.

The first step is to form Routh's array, and an undamped oscillation will exist in the second row if s^1 row equals zero.

$$
\begin{array}{ccc}
s^3 & 1 & c \\
s^2 & b & Kd \\
s^1 & \dfrac{bc - Kd}{b} & \\
s^0 & Kd &
\end{array}
$$

The auxiliary equation is

$$bs^2 + Kd = 0$$

Where the solution is

$$s_{1,2} = \sqrt{\frac{-Kd}{b}} = \pm j \sqrt{\frac{Kd}{b}} = \pm j\omega_n$$

To find the value of K, set $s^1 = 0$; then:

$$s^1 = 0 = \frac{bc - Kd}{b} = bc - Kd$$

$$Kd = bc \text{ and } K = bc/d$$

Then, substituting K into $\pm j\omega$ yields the following solution.

$$s_{1,2} = \pm j \sqrt{\frac{Kd}{b}} = \pm j \sqrt{\left(\frac{bc}{d}\right)\left(\frac{d}{b}\right)} = \pm j \sqrt{c}$$

Thus, it can be concluded that $K = bc/d$ is the value of the static sensitivity, K, at crossover and is stable in the range $0 < K < bc/d$.

6.5   SUMMARY OF ROOT LOCUS PROCEDURES

1. Derive open-loop transfer function.
2. Factor numerator and denominator.
3. Plot zeros and poles in the s plane.
4. Find out where locus exists on real axis.
5. Determine asymptotes as $s \leftarrow \infty$.
6. Select the search point; remember that it must satisfy angular condition.
7. Determine the static loop sensitivity that must satisfy magnitude condition

$$K = \frac{\prod(\text{dir segment from poles})}{\prod(\text{dir segment from zeros})}$$

8. Determine the roots at the desired operating point.
9. Take the inverse Laplace transform.
10. Plot the time response to determine the desired specifications.
11. Compensate if necessary.

6.6   ADDITION OF POLES AND ZEROS

Let us look qualitatively at the root locus and the effects of adding a pole or a zero to the locus of the following equation with static loop sensitivity $K > K\sigma$.

$$G(s) = \frac{K}{s\left(s + \dfrac{1}{T_1}\right)}$$

The root locus for the equation has poles at 0 and $1/T_1$, and breaks away halfway between the two poles.

If a zero is added to the system on the real axis at the location $-1/T_2$ such that the zero lies to the left of the $-1/T_1$ pole, the locus will be pulled to the left farther away from the $j\omega$ axis. The locus that will have a break-in at a point less than $-1/T_2$ on branch will go to $-\infty$, and the other branch will terminate at the +0; hence, the roots will be farther to the left, away from the $\pm j\omega$ axis. Next, it should be noted that the farther the roots are from the imaginary axis, the larger is the absolute magnitude σ and the faster is the decay of any transients.

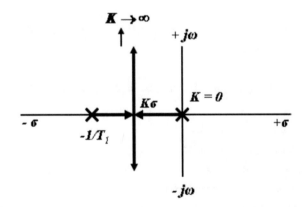

If a pole is added to the system on the real axis of the root locus at the location $-1/T_3$ such that the pole lies to the left side of the $-1/T_1$ pole, the pole will push the locus to the right so that the branches cross the $j\omega$ axis. The closer the roots are to the imaginary axis, the smaller is the absolute magnitude σ and the slower is the decay of any transients.

In summary, the addition of a zero has the effect of pulling root locus toward the left, thereby tending to make the system "more" stable and faster to respond. On the other hand, the addition of a pole has the effect of pulling root locus toward the right, thereby tending to make the system a "less" stable and slower to respond.

REFERENCES

[1]   D'Azzo, J. J., and Houpis, C. H., *Feedback Control System Analysis and Synthesis*, 2nd ed., McGraw-Hill Book Company, New York (1966).

[2]   Lewis, P. H., and Yang, C., *Basic Control Systems Engineering*, Prentice-Hall, Inc., Upper Saddle River, NJ (1997).

[3]   Phillips, C. L., and Harbor, R. D., *Feedback Control Systems*, 4th ed., Upper Saddle River, NJ (2000).

CHAPTER 7

Frequency Response Analysis

7.1 STEADY-STATE FREQUENCY RESPONSE

In the past, steady-state frequency response was determined after the root locus was determined and the system gain was set for desired performance. Frequency response analysis is the second basic method for predicting and adjusting system performance without resorting to the actual solving of the system's differential equation.

The frequency response of a control system is defined as curves or plots on semilog graph paper of the following characteristic curves:

1. The absolute value of the magnitude (M) of the control ratio $C(j\omega)/R(j\omega)$ plotted against the frequency ω;
2. The phase angle α of the control ratio $C(j\omega)/R(j\omega)$ plotted against the frequency ω.

Frequency response curves are often called the Bode plots. The two curves present a qualitative picture of the system's transient response and enable the designer to minimize noise within the system [1, 2].

7.2 FIGURES OF MERIT USED TO MEASURE SYSTEM PERFORMANCE

As discussed in a previous chapter, system performance is generally judged on the following figures of merit:

1. Peak overshoot, M_p, or M_m, which is the amplitude of the first overshoot.
2. Peak time, T_p, which is the time to reach the maximum peak overshoot.
3. Settling time, T_s, is the time it takes for the response to first reach and thereafter remain within 2% of the final value, which occurs at about four time constants ($t = 4T$) and $T_s = 4/\zeta\omega_n$.
4. The number of oscillations, N, in the response up to settling time, or the damped frequency of oscillation of the transient (ω_d) [1].

7.3   RELATIONSHIP BETWEEN THE ROOT LOCUS AND THE FREQUENCY RESPONSE

Recall that the root locus method incorporates the most desirable features of both the classical method and the steady-state sinusoidal phasor analysis. Figures 7.1–7.6 show the plots of the root locus in the s lane and their correlation with the transient solutions of the frequency response in the time domain for a step input.

Figure 7.1a shows the root locus for a second-order, underdamped system that is stable with a pair of complex conjugate roots and a damping factor (ζ) of less than 1 ($0 < \zeta < 1$). Figure 7.1b is the corresponding time-domain, underdamped frequency response to a step function [1].

Figure 7.2a shows the root locus for a second-order, critically damped system that is stable with a pair of real roots and a damping factor equal to one ($\zeta = 1$). Figure 7.2b is the corresponding time-domain, critically damped frequency response to a step function [2–4].

Figure 7.3a shows the root locus for a second-order, overdamped system that is stable with a pair of real roots and a damping factor greater than 1 ($\zeta > 1$). Figure 7.3b is the corresponding time-domain, overdamped frequency response to a step function.

Figure 7.4a shows the root locus for a second-order, underdamped, sustained oscillation system that is stable with a pair of imaginary roots on the $j\omega$ axis and a damping factor equal to zero ($\zeta = 0$). Figure 7.4b is the corresponding time-domain, underdamped, sustained oscillation frequency response to a step function [1, 3, 4].

Figure 7.5a shows the root locus for a second-order system that is unstable with a pair of real roots on the right side of the s plane and a damping factor less than zero ($\zeta < 0$). Figure 7.5b is the corresponding time domain for the unstable system's frequency response to a step function [1, 3, 4].

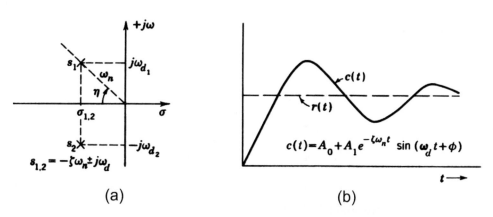

(a)   (b)

FIGURE 7.1: Plots of the root locus and its corresponding underdamped frequency response to a step function.

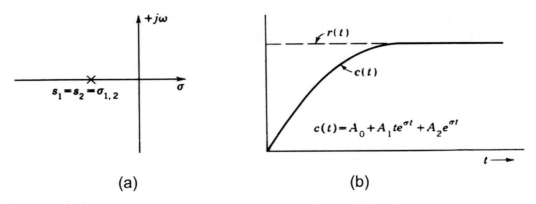

(a)   (b)

FIGURE 7.2: Plots of the root locus and its corresponding critically damped response to a step function.

Figure 7.6a shows the root locus for a second-order system that is unstable with a pair of complex, conjugate roots on the right side of the s plane and a damping factor (ζ) less than zero ($\zeta < 0$). Figure 7.6b is the corresponding time domain for the unstable system's unbound frequency response to a step function [1].

7.4   CONSTANT PARAMETERS ON S PLANE

Let us review a second-order system with control ratio whose transient response is as follows:

$$\frac{C(s)}{R(s)} = \frac{K}{s^2 + 2\zeta\omega_n s + \omega_n^2}$$

The time domain response has the general form:

$$c(t) = C_1 e^{s_1 t} + C_2 e^{s_2 t}$$

where

$$s_1 = -\zeta\omega_n + j\omega_n\sqrt{1 - \zeta^2} = \sigma + j\omega_d$$

$$s_2 = -\zeta\omega_n - j\omega_n\sqrt{1 - \zeta^2} = \sigma - j\omega_d$$

For $\zeta < 1$, and $c(t) = Ae^s \sin(\omega_d t + \varphi)$, the radius r in the s plane is equal to the undamped natural frequency (ω_n) of the system and calculated as follows.

$$r = \sqrt{\omega_{d_1}^2 + \sigma_1^2} = \sqrt{\omega_n^2(1 - \zeta^2) + \omega_n^2\zeta^2} = \omega_n$$

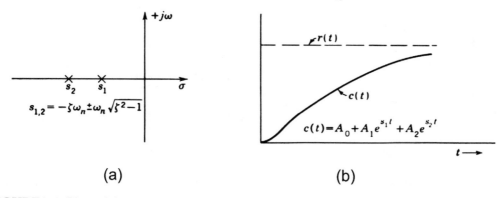

(a)   (b)

FIGURE 7.3: Plots of the root locus and its corresponding overdamped response to a step function.

Because the cosine of η is the damping factor as shown in the following equation, the inverse must be that the angle η is the inverse cosine of the damping factor.

$$\cos\eta = \left|\frac{-\sigma_1}{r}\right| = \frac{\zeta\omega_n}{\omega_n} = \zeta$$

Therefore,

$$\eta = \cos^{-1}\zeta$$

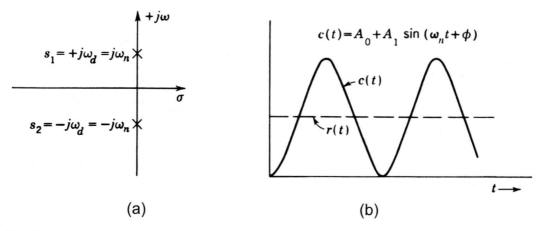

(a)   (b)

FIGURE 7.4: Plots of the root locus and its corresponding underdamped, sustained oscillation frequency response to a step function [1].

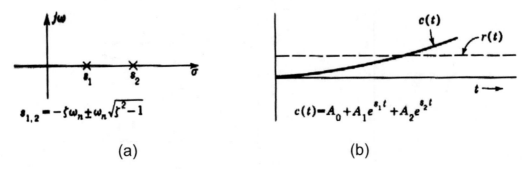

(a)   (b)

FIGURE 7.5: Plots of the root locus and its corresponding unstable system with a pair of real roots on the right side of the s plane response to a step function.

The angle η is measured clockwise from the negative real axis for positive damping factor (ratio) ζ. The damping factor ζ is the cosine of the angle η. Figure 7.7 shows the constant parameters of the s plane.

Note in Figure 7.7 that the horizontal lines indicate constant values of the damped natural frequency, ω_d, and that the vertical lines represent lines of constant damping, σ. The circular lines about the origin are circles of constant undamped natural angular frequencies, whereas the radial lines passing through the origin at angle η are lines of constant damping ratio, ζ.

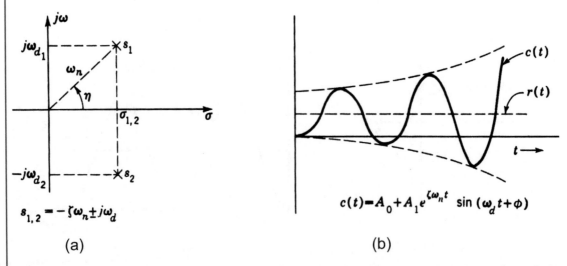

(a)   (b)

FIGURE 7.6: Plots of the root locus and its corresponding unstable system with a pair of complex, conjugate roots on the right side of the s plane response to a step function.

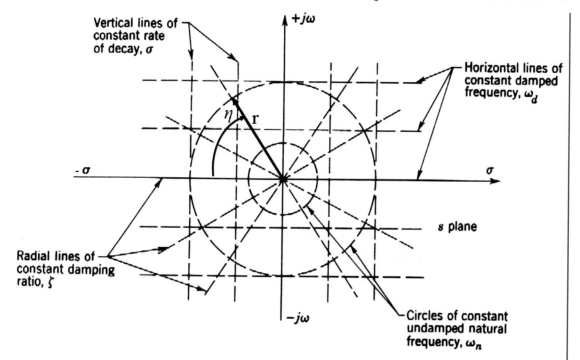

FIGURE 7.7: Constant parameters on s plane.

The key points to remember are the types of roots that are associated with the damping ratio as shown in Table 7.1 and Figure 7.8.

The following equations are used to calculate the peak of the log magnitude, M_{m}, and the resonant frequency for the damping ratio, $\zeta < 1$.

TABLE 7.1: Roots versus damping in s plane	
DAMPING	ROOTS
$\zeta = 1$	s_{12} real and repeated
$\zeta > 1$	s_{12} real and equal
$\zeta < 1$	s_{12} complex and conjugate
$\zeta = 0$	$s_{12} = j\omega_n$ and imaginary

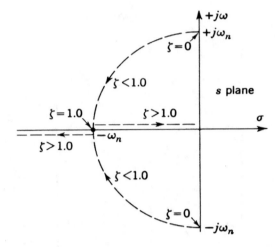

FIGURE 7.8: Roots versus damping in s plane.

$$M_{\mathrm{m}} = \frac{1}{2\zeta\sqrt{(1-\zeta^2)}} \text{ and}$$

$$\omega_{\mathrm{m}} = \omega_{\mathrm{n}}\sqrt{(1-2\zeta^2)} \text{ for } \omega_{\mathrm{m}} = \text{ real value}$$

Both parameters are dependent on the damping ratio, ζ, and whenever the damping factor is less that 0.707 ($\zeta < 0.707$), M_{m} is greater than 1.

Note that for the calculation of the damped natural frequency, the equation is as follows:

$$\omega_{\mathrm{d}} = \omega_{\mathrm{n}}\sqrt{1-\zeta^2}$$

7.5   DRAWING THE BODE PLOTS

From the transfer function of the control ratio, the log magnitude and phase diagram are drawn on semilog graph paper, because in the logarithm domain:

1. The arithmetic operations of multiplication and division become addition and subtraction, respectively.
2. The basic factors of the transfer function fall into four categories, which may be plotted by means of straight-line asymptotic approximations.
3. The use of semilog paper eliminates the need to take the logarithms of the frequency values.

4. Finally, the semilog paper expands the low frequency range, which is often of primary importance in system design [1,2].

In mathematics, recall that the natural logarithm (ln) to the exponential base (e) of a complex number is also complex as shown in the following equation.

$$\ln|G(j\omega)|\, e^{j\phi(\omega)} = \ln|G(j\omega)| + \ln e^{j\phi(\omega)}$$
$$= \ln|G(j\omega)| + j\phi(\omega)$$

Likewise, the logarithm to base 10 of a complex number is a complex as shown in the following equation. Note that the phase (φ) is adjusted by a factor of 0.434.

$$\log|G(j\omega)|\, e^{j\phi(\omega)} = \log|G(j\omega)| + \log e^{j\phi(\omega)}$$
$$= \log|G(j\omega)| + j0.434\phi(\omega)$$

In control or feedback system analysis of the log magnitude, the units most commonly used is decibel. The term "decibel" was originally used in communications engineering when referring to the ratio of two values of power. It is more common to use the ratio of two voltages instead of power as shown in the following equation.

$$\frac{P_1}{P_2} = \frac{\dfrac{V_1^2}{R}}{\dfrac{V_2^2}{R}} = \frac{V_1^2}{V_2^2} = \left(\frac{V_1}{V_2}\right)^2$$

The decibel is a dimensionless quantity that was 10 times the log of the ratio of two values of power. In terms of voltages, the decibel is 20 times the log of the ratio of two values of voltages; hence, the log magnitude of the control ratio is:

$$Lm G(j\omega) = 20\log|G(j\omega)|\, dB$$

Control and communication engineers used the terms "octave" and "decade" to express ratios of frequency, for example,

1. Octaves: $f_2/f_1 = 2$ or $f_2 = 2f_1$
2. Decades: $f_2/f_1 = 10$ or $f_2 = 10f_1$

Table 7.2 shows the decibel values of common numbers. Note that as a number doubles, the increase in decibels is 6 dB.

TABLE 7.2: Decibel values of common numbers

NUMBER	DECIBELS
0.01	−40
0.1	−20
0.5	−6
1	0
2	6
10	20
100	40
200	46

7.6   FACTORS IN LOG MAGNITUDE

Transfer functions (numerator and denominator) have four basic factors:

1. Constants (K_n): The equation for the system gain is $L_mK_n = 20 \log K_n$, which is a horizontal straight line.
2. The factor $j\omega$ or s term in the denominator is $L_m(1/j\omega) = 20 \log|1/j\omega| = -20 \log(\omega)$ when plotted against frequency. The $j\omega$ term is drawn as a straight line with a negative slope of 6 dB/octave or 20 dB/decibel. The phase angle for the s or $j\omega$ term is constant and equal to $-90°$.

   The equation for the factor $j\omega$ in the numerator is $L_m(1/j\omega) = 20 \log|1/j\omega| = +20 \log(\omega)$ that, when plotted against frequency, is a straight line with a positive slope of 6 dB/octave or 20 dB/decibel. The phase angle for s or $j\omega$ in the denominator is constant and equal to $+90°$.
3. The equation for the factor $(1 + j\omega T)$ in the denominator is:

$$L_m\frac{1}{1 + j\omega T} = 20 \log\left|\frac{1}{1 + j\omega T}\right| = -20 \log \sqrt{1 + j\omega^2 T^2}$$

The asymptotes of the plot of $L_m[1/(1 + j\omega T)]$ are two straight lines: one of zero slope for values of ω less than $\omega = 1/T$, and the other one of -6 dB/octave slope for values of ω greater than $\omega = 1/T$. The frequency at which the asymptote lines intersect is defined as the corner frequency, $\omega_{cf} = 1/T$. The exact values of the log magnitude curve for the factor $(1 + j\omega T)$ in the denominator are shown in Table 7.3 [1].

Next, draw the phase diagram and start by plotting the following five points, shown in Table 7.4, and connect the points with straight lines [1].

From point $-5.7°$ extend the line to $0°$, and from point $-84.3°$ extend to $-90°$.

An example of the log magnitude and phase diagram for the equation $(1 + j\omega T)^{-1}$ or $[1 + j(\omega/\omega_d)]^{-1}$ is shown in Figure 7.9.

4. The quadratic factors have the form:

$$\left[1 + \frac{2\zeta}{\omega_n}j\omega + \frac{1}{\omega_n^2}(j\omega)^2\right]^{-1}$$

For $\zeta > 1$, factor into two first-order factors with real poles; the same is true for $\zeta = 1$. For $\zeta < 1$, the factors are complex and conjugate, which can be plotted without factoring:

$$L_m\left[1 + \frac{2\zeta}{\omega_n}j\omega + \frac{1}{\omega_n^2}(j\omega)^2\right]^{-1} = -20\log\left[\left(1 - \frac{\omega^2}{\omega_n^2}\right)^2 + \left(\frac{2\zeta\omega}{\omega_n}\right)^2\right]^{\frac{1}{2}}$$

$$\text{Angle}\left[1 + \frac{2\zeta}{\omega_n}j\omega + \frac{1}{\omega_n^2}(j\omega)^2\right]^{-1} = -\tan^{-1}\left(\frac{\dfrac{2\zeta\omega}{\omega_n}}{1 - \dfrac{\omega^2}{\omega_n^2}}\right)$$

TABLE 7.3: Exact values of the log magnitude curve for the factor $(1 + j\omega T)$

LOCATION	VALUE, dB	PHASE ANGLE
At ω_{cf}	-3	$-45°$
One octave	-1	$-63.4°$
Two octaves	-0.26	
One decade	-0.04	$-90°$

TABLE 7.4: Key phase angle points		
RATIO (ω/ω_{cf})	**RANGE FROM ω_{cf}**	**PHASE ANGLE FROM $0°$**
0.1	−Decade	−5.7°
0.5	−Octave	−26.6°
1	At ω_{cf}	−45°
2	Octave	−63.4°
10	Decade	−84.3°

For large values of frequency ω, the log magnitude is approximately $-40\log(\omega/\omega_n)$, which is a line with slope of −40 dB/decade. The corner frequency is equal to the undamped natural frequency of the quadratic term $\omega = \omega_n$, and resonance is in the vicinity of $\omega = \omega_n$.

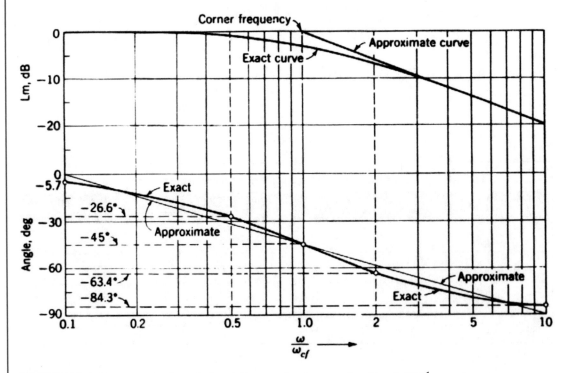

FIGURE 7.9: Log magnitude and phase diagram for the equation $(1 + j\omega T)^{-1}$.

A family of log magnitude and phase curves for several values of $\zeta < 1$ maybe used to plot the two curves. In addition, the control engineer may use the family of curves for several values of $\zeta < 1$ that are based on corrections of each actual curve from the straight line asymptotes [2,5].

7.7 DERIVING THE TRANSFER FUNCTION FROM THE LOG MAGNITUDE CURVES

Often, the engineer's task involves collecting stimulus-response data from a system and then deriving the system transfer function; this is particularly true for physiological systems. For any given log magnitude curve, the system type and gain can be determined. Let us examine a few examples. Figure 7.10 shows the log magnitude plot for a type 0 system.

The slope at low frequencies of a type 0 system is zero, whereas its magnitude at low frequencies is $20 \log K_0$, where the gain K_0 is the static step error coefficient [1].

Figure 7.11 shows the characteristic log magnitude curves for two type 1 systems that the slope of the curve at low frequencies is -20 dB/decade. The intercept of the low-frequency slope with the 0-dB axis occurs at frequency ω_x, where $\omega_x = K_1$ and the gain K_1 is the static parabolic error coefficient. The value on the low-frequency slope at $\omega = 1$ is equal to $20 \log K_1$ [1].

Figure 7.12 shows the characteristic log magnitude curves for two type 2 systems that the slope of the curve at low frequencies is -40 dB/decade. The intercept of the low-frequency slope with the 0-dB axis occurs at a frequency ω_y, where $\omega_y^2 = K_2$ and the gain K_2 is the static ramp error coefficient. The value on the low-frequency slope at $\omega = 1$ is equal to $20 \log K_2$ [1].

7.8 SUMMARY

In summary, the procedures used in developing frequency response graphs are as follows.

1. Derive the open-loop transfer function $G(s)H(s)$ of the system.
2. Arrange the factors of the transfer function $G(s)H(s)$ so that they are in the form: $j\omega$, $(1 + j\omega)$, and $[1 + aj\omega + b(j\omega)2]$.

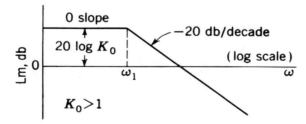

FIGURE 7.10: Log magnitude for a type 0 system.

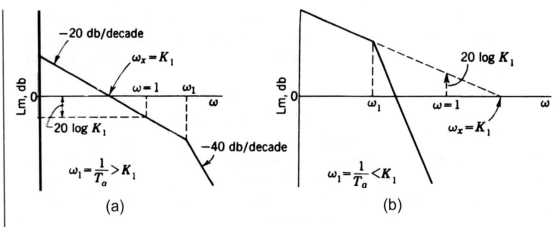

FIGURE 7.11: Log magnitude curves for two type 1 systems.

3. Plot either the approximate or the exact log magnitude in the phase angle diagrams for the transfer function $G(s)H(s)$.

4. Transfer the data from the log magnitude in the phase angle plots to a Nichols plot, which combines the log magnitude versus the phase angle in a single graph.

5. Apply the Nyquist stability criterion and adjust the gain for the desired degree of stability, M_m, of the system. Check the correlation of the time domain output response for a step input signal.

6. If the qualitative response does not meet the desired specification, determine the shape that the plot must have to meet specifications.

7. In the final step, synthesize the compensator that must be inserted into the system, if gain adjustment is not enough.

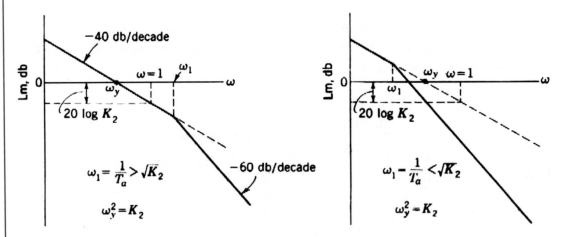

FIGURE 7.12: Log magnitude curves for two type 2 systems.

REFERENCES

[1]   Chestnut, H., and Mayer, R. W., *Servomechanisms and Regulating Systems*, John Wiley & Sons, New York, NY (1954).

[2]   Van Valkenburg, M. E., *Network Analysis*, 2nd ed., Prentice-Hall, Inc., Englewood Cliffs, NJ (1964).

[3]   Lewis, P. H., and Yang, C., *Basic Control Systems Engineering*, Prentice-Hall, Inc., Upper Saddle River, NJ (1997).

[4]   Phillips, C. L., and Harbor, R. D., *Feedback Control Systems*, 4th ed., Upper Saddle River, NJ (2000).

[5]   D'Azzo, J. J., and Houpis, C. H., *Feedback Control System Analysis and Synthesis*, 2nd ed., McGraw-Hill Book Company, New York (1966).

· · · ·

CHAPTER 8

Stability and Margins

A system's stability and degree of stability can be determined from the *log magnitude* and *phase diagram*. The stability characteristics are specified in terms of the following quantities [1–4]:

1. Gain crossover—the point where the magnitude is unity [1] or $L_m G(j\omega) = 0$ dB. The frequency at gain crossover is called the "phase-margin frequency" (ω_φ). Do not confuse phase margin frequency, ω_φ, with phase margin, γ.
2. Phase margin—180° plus the angle of the transfer function φ at the gain crossover point. Phase margin is designated as:

$$\gamma = 180° + \varphi,$$

where φ is negative.

   The phase margin, γ, is the amount of phase shift φ at the frequency ω_φ that would just produce instability.

3. Phase crossover—the point of the transfer function at which the phase angle is −180°; the frequency at which the phase crossover occurs is called *gain margin frequency*, ω_c.
4. Gain margin—the additional gain α that just makes the system unstable. Expressed in terms of transfer function at the frequency ω_c, it is:

$$|G(j\omega)|\alpha = 1 \text{ and } |G(j\omega)| = 1/\alpha$$

or in terms of log magnitude, $L_m \alpha = -L_m G(j\omega)$ dB, which identifies the gain margin on the log magnitude diagram.

In general, for "minimum-phase networks," which consist of poles and no zeros, the phase margin must be positive for a stable system. A negative phase margin means that the system is unstable. Recall that the phase margin is related to the effective damping ratio, ζ, of the system. Angles between 45° and 60° produce a satisfactory system response.

The phase angle, φ, of the transfer function at gain crossover must be more than the −180° point.

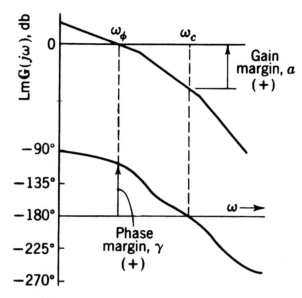

FIGURE 8.1: Example of stable log magnitude margins.

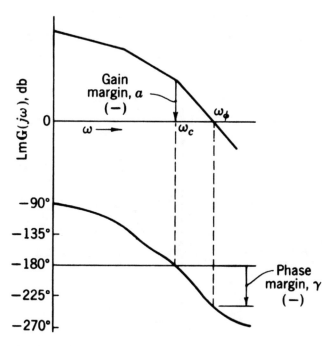

FIGURE 8.2: Example of unstable log magnitude margins.

The gain margin must be positive, when expressed in decibels (dB) for the system to be stable.

Recall that gain margin is the amount of gain that can be added before the system becomes unstable. A negative gain margin means that the system is unstable. The slope at gain crossover must be more positive than −40 dB [1,2].

Let us examine two examples, the first in which the system is stable and the other in which the system is unstable. Figure 8.1 shows an example of "stable" log magnitude margins. Note that in Figure 8.1, both the phase margin, γ, and gain margin, α, are positive; therefore, the system is stable.

Figure 8.2 shows an example of "unstable" log magnitude margins. Note that in Figure 8.2, both the phase margin, γ, and gain margin, α, are negative; therefore, the system is unstable.

8.1 NICHOLS CHARTS

A Nichols chart is a single plot of the log magnitude versus phase angle, where the origin of the Nichols chart is 0 dB gain and −180° phase angle. Figure 8.3 shows an example of a Nichols chart where the phase margin, γ, is displacement from the y-axis to the $G(s)H(s)$ trace at gain crossover frequency, ω_φ, and the gain margin is displacement from the $G(s)H(s)$ curve to the x-axis at phase crossover, ω_c [1–4].

Note that increasing the gain will increase the curve of the transfer function, $G(s)H(s)$; thus, decreasing the gain margin will degrade the stability of the system. If the gain is increased so that

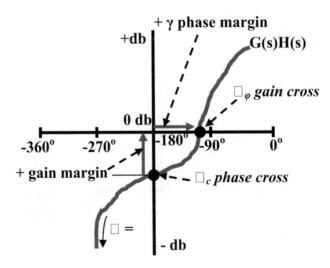

FIGURE 8.3: Example of a Nichols chart.

the transfer function has a positive log magnitude value at −180°, this will result in a negative gain margin and/or a negative phase margin, which means that the system is unstable.

REFERENCES

[1]   Lewis, P. H., and Yang, C., *Basic Control Systems Engineering*, Prentice-Hall, Inc., Upper Saddle River, NJ (1997).

[2]   Chestnut, H., and Mayer, R. W., *Servomechanisms and Regulating Systems*, John Wiley & Sons, Inc., New York, NY (1954).

[3]   Phillips, C. L., and Harbor, R. D., *Feedback Control Systems*, 4th ed., Upper Saddle River, NJ (2000).

[4]   D'Azzo, J. J., and Houpis, C. H., *Feedback Control System Analysis and Synthesis*, 2nd ed., McGraw-Hill Book Company, New York (1966).

· · · ·

CHAPTER 9

Introduction to LabVIEW

9.1   WHAT IS LABVIEW?

Laboratory Virtual Instrumentation Engineering Workbench (LabVIEW) is a graphical programming environment used for test, measurement, and control applications. It uses icons instead of lines of text to create applications, which is in contrast to text-based programming languages, where instructions determine program execution. LabVIEW programs are called virtual instruments (VIs) because their appearance and operation imitate physical instruments.

9.2   ENVIRONMENT

9.2.1   Getting Started

You can launch LabVIEW from Start ≫ Programs ≫ National Instruments ≫ LabVIEW, which opens the "Getting Started" window (Figure 9.1). From here, you can create new VIs and projects, open existing ones, find examples, and search the LabVIEW Help. Once you create or open a VI, this window closes.

To create a new VI, click on "Blank VI" under "New." You can open a VI by either choosing a VI from the list shown under "Open," or by choosing to browse to the file that you want.

9.2.2   Front Panel

The front panel window (Figure 9.2) is the user interface for the VI. This consists of controls and indicators, which are the interactive input and output terminals of the VI, respectively. You will learn how to insert them on the front panel in the section on Functions/Controls palette.

9.2.3   Block Diagram

The block diagram (Figure 9.3) is the window that displays the "code" behind the LabVIEW VI. Block diagram objects include terminals, sub-VIs, functions, constants, structures, and wires, which transfer data among other block diagram objects. You will learn how to insert them on the block diagram in the section on Functions/Controls palette.

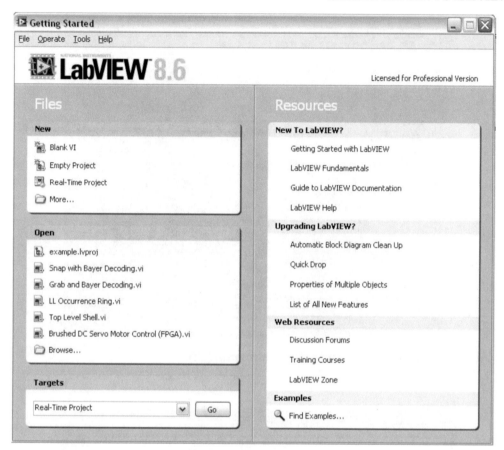

FIGURE 9.1: LabVIEW Getting Started window.

9.2.4   Controls and Indicators

Controls and indicators behave as inputs and outputs of the block diagram algorithm. On the front panel, they can be seen as shown Figure 9.4. On the block diagram, they appear as terminals (Figure 9.5).

Tip: For any object in the block diagram that has input and output terminals, you can create a control/constant (Figure 9.6) for the input and an indicator for the output by following these steps:

1.   Right click on the input/output terminal.
2.   From the pop-up menu, select Create ≫ Control/Constant/Indicator.

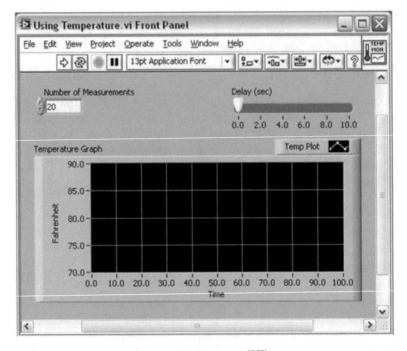

FIGURE 9.2: Front panel window of a virtual instrument (VI).

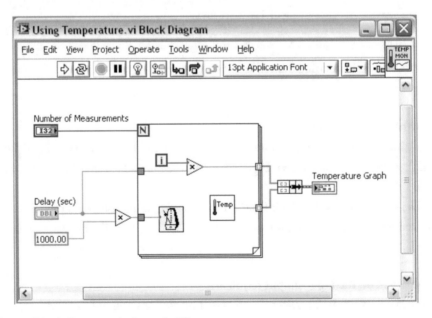

FIGURE 9.3: Block diagram window of a VI.

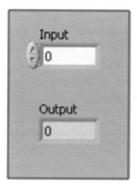

FIGURE 9.4: Control and indicator on the front panel.

9.2.5   Functions/Controls Palette

The Functions palette (Figure 9.8) contains the VIs, functions, and constants that are used to cre-ate the block diagram. You can access the Functions palette from the block diagram by selecting View ≫ Functions Palette or by right-clicking anywhere in the block diagram. To insert a block diagram object, click on it in the Functions palette and then click on the space in the block diagram where you want it inserted.

The Controls palette (Figure 9.7) contains the controls and indicators used to create the front panel. You can access the Controls palette from the front panel window by selecting View ≫ Controls Palette or by right-clicking anywhere in the front panel. To insert a control or indicator on the front panel, click on the desired object in the palette and then click on where you want it inserted on the front panel.

9.3   VIRTUAL INSTRUMENTS

9.3.1   Data Flow Execution

LabVIEW follows a dataflow model for running VIs (Figure 9.9). A block diagram node executes when it receives all required inputs. When a node executes, it produces output data and passes the data to the next node in the dataflow path. The movement of data through the nodes determines the execution order of the VIs and functions on the block diagram.

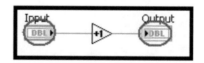

FIGURE 9.5: Control and indicator on the block diagram.

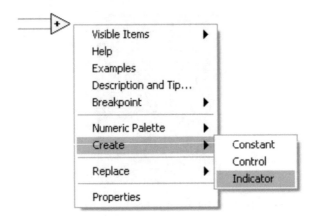

FIGURE 9.6: Creating a control, constant, or indicator.

Two important points to keep in mind are:

1. A node executes only when data are available at all of its input terminals.
2. A node supplies data to the output terminals only when the node finishes execution.

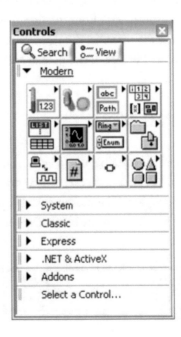

FIGURE 9.7: Controls palette.

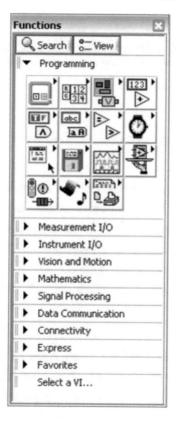

FIGURE 9.8: Functions palette.

9.3.2 Running a VI

You can run the VI by clicking on the Run button (Figure 9.10), which appears as a white arrow on the toolbar in the front panel or block diagram of the VI.

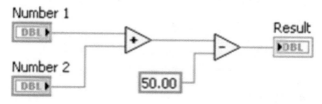

FIGURE 9.9: Dataflow in LabVIEW.

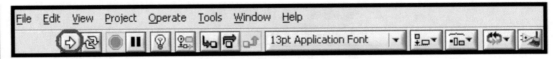

FIGURE 9.10: Run button on the toolbar of a VI.

9.4 LABVIEW RESOURCES
9.4.1 Example Finder
From the "Getting Started" window, you can open the Example Finder by clicking on "Find Examples" on the bottom right. If a VI is already open, choose Help >> Find Examples to open it. You can use this utility to browse or search examples installed on your computer or on the NI Developer Zone at ni.com/zone. This utility is an extremely useful way of learning how to create a VI for your application and can even be modified for this purpose.

Have an example here to direct to "Bouncing Ball 3D" VI?

9.4.2 Context Help
In the front panel or block diagram, there is a question mark icon that is located in the top right portion. Clicking on this brings up the Context Help window (Figure 9.11), which provides a brief description of any block diagram object that the mouse pointer hovers over. For example, the context help window displayed above is shown when the pointer is over the Add function in the block diagram. Click on Detailed help at the bottom of the description for detailed information.

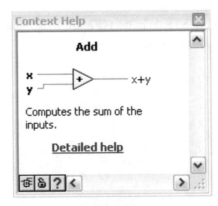

FIGURE 9.11: Context Help window.

9.4.3 LabVIEW Help

You can access LabVIEW Help from the front panel or block diagram window by selecting Help ≫ Search the LabVIEW Help. This help system includes information about LabVIEW programming concepts, step-by-step instructions for using LabVIEW, and reference information about LabVIEW VIs, functions, palettes, menus, and tools.

9.5 STRUCTURES/PROGRAMMING CONSTRUCTS

9.5.1 While Loops

The While Loop (Figure 9.12) is a structure that executes the code within it until a condition occurs. It is located on the Structures palette within the Functions Palette (Programming ≫ Structures ≫ While Loop). Select the While Loop from the palette, then use the cursor to drag a selection rectangle around the section of the block diagram you want to repeat. When you release the mouse button, a While Loop boundary encloses the section you selected. Add block diagram objects to the While Loop by dragging and dropping them inside the While Loop.

A While Loop executes the subdiagram until the conditional terminal, an input terminal, receives a specific Boolean value. It runs infinitely if the condition never occurs and has to execute at least once.

9.5.2 For Loops

The For Loop (Figure 9.13) is a structure that executes the code within it a set number of times. It is located on the Structures palette as well. It is placed on the block diagram just like a While Loop. It has a count terminal and an iteration terminal. The count terminal is an input terminal whose value indicates how many times to repeat the subdiagram. The iteration terminal is an output terminal

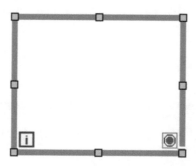

FIGURE 9.12: While loop.

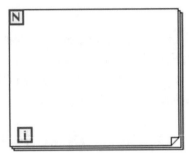

FIGURE 9.13: For loop.

that contains the number of completed iterations. Unlike a While Loop, a For Loop does not have to execute at all if the count terminal is set to zero.

9.5.3   MathScript Node

The Mathscript node (Figure 9.14) is found on the Structures palette within the Functions Palette. It executes LabVIEW MathScripts. You can use the MathScript Node to evaluate scripts that you create in the LabVIEW MathScript Window. The LabVIEW MathScript syntax is similar to the MATLAB® language syntax.

Input and output terminals are inserted in order to pass data between the script within the node and your VI. You can create an input or output terminal by right-clicking on the border of the node and selecting "Add Input" or "Add Output," respectively. Always make sure that the data type of the terminal matches the data type of the input or output value that you want. You can choose the data type by right-clicking on the terminal and selecting "Choose Data Type." For example, in case you have specified that the input/output value is an array of doubles in the script within the node, you can right-click on the respective terminal and choose the data type as 1D-array ≫ DBL 1D.

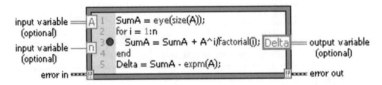

FIGURE 9.14: MathScript node.

9.6   DATA STRUCTURES

9.6.1   Constants

Constants are terminals on the block diagram that supply fixed data values to the block diagram. They will not show up on the front panel of the VI.

9.6.2   Arrays

An array (Figure 9.15) consists of elements and dimensions. Elements are the data that make up the array. A dimension is the length, height, or depth of an array. Every element in an array has an index number associated with it. Arrays are zero-indexed, which implies that the first element has index zero.

You can build arrays of numeric, Boolean, path, string, waveform, and cluster data types. Consider using arrays when you work with a collection of similar data and when you perform repetitive computations. They are ideal for storing data collected from waveforms or data generated in loops, where each iteration of a loop produces one element of the array.

An array shell can be found in the Controls Palette under "Array, Matrix & Cluster" (Controls ≫ Modern ≫ Array, Matrix & Cluster ≫ Array). An array control or indicator can be created on the front panel by placing the array shell and dragging a data object or element into the shell.

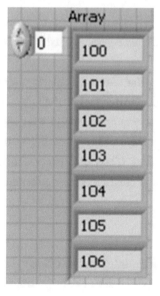

FIGURE 9.15: Array control on the front panel.

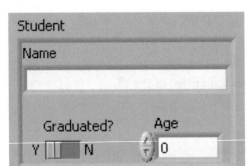

FIGURE 9.16: Cluster control on the front panel.

You must insert an object in the array shell before you use the array on the block diagram. Otherwise, the array terminal appears black with an empty bracket and has no data type associated with it.

9.6.3   Clusters

Clusters (Figure 9.16) group data elements of mixed types. Bundling several data elements into clusters eliminates wire clutter on the block diagram.

Although cluster and array elements are both ordered, you must unbundle all cluster elements at once or use the Unbundle By Name function to access specific cluster elements. Clusters also differ from arrays in that they are fixed size. A cluster cannot contain a mixture of controls and indicators.

Cluster shells can be found in the Controls Palette under "Array, Matrix & Cluster" (Controls ≫ Modern ≫ Array, Matrix & Cluster ≫ Cluster), just like array shells. A cluster control is created in a similar manner to an array control, except that you can drag objects of different data types into the shell on the front panel. A cluster shell constant can be found in the Functions Palette under Programming ≫ Cluster, Class & Variant. In this shell, you can place string, numeric, Boolean, or cluster constants to create a cluster constant. This can be used to store constant data.

9.7   GRAPHS AND CHARTS

Waveform graphs and charts are commonly used to display data on the front panel in LabVIEW. These indicators facilitate the display of a set of data points for measurement and analysis. These points can be plotted one data point at a time or as a set of acquired data points. These indicators can be found in the controls palette under Controls ≫ Modern ≫ Graph.

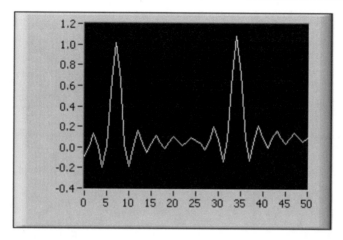

FIGURE 9.17: Waveform graph.

9.7.1 Waveform Graph

Waveform graphs are usually used when an array of points are to be plotted. They can be used to graph one or more plots of evenly spaced measurements. Figure 9.17 shows what a waveform graph looks like.

9.7.2 Waveform Chart

Waveform chart is a special type of numeric indicator that displays one or more plots of data typically acquired at a constant rate. They can display one or more plots just like waveform graphs.

Waveform charts can plot data in three different modes:

1. Strip mode
   This mode has a scrolling display that is similar to a paper tape strip chart recorder. As it receives each new value, it plots the value at the right margin, and shifts old values to the left.
2. Scope mode
   This mode has a retracing display similar to an oscilloscope. As it receives each new value, it plots the value to the right of the last value. When the plot reaches the right border of the plotting area, it erases the plot and begins plotting again from the left border.
3. Sweep mode
   This mode acts much like the scope chart, but the plot is not erased when the plot hits the right border. Instead, a moving vertical line marks the beginning of new data and moves across the display from left to right as it adds new data.

Figure 9.18 shows what a waveform chart looks like:

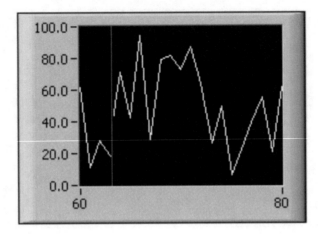

FIGURE 9.18: Waveform chart.

9.8   WHAT IS THE DIFFERENCE?

Waveform charts (Figure 9.19) are mainly used when data points are plotted on the fly. Thus, they would be very useful when the chart has to be updated with new points at a constant rate.

Waveform graphs, on the other hand, are useful when displaying an accumulated set of points on the front panel. In these cases, all the points are acquired first and then are displayed together on the front panel. In the block diagram window, waveform graphs (Figure 9.20) appear similar to waveform charts (Figure 9.19), therefore keep the owned label of the waveform nodes visible.

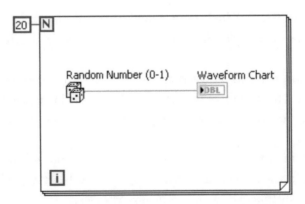

FIGURE 9.19: Waveform chart in a block diagram.

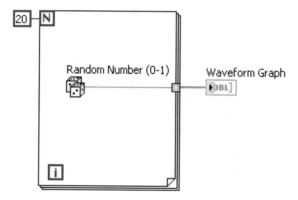

FIGURE 9.20: Waveform graph in a block diagram.

9.9   SUMMARY

LabVIEW is a powerful tool that permits the user to design any type of VI for gathering sensor or image information, analyzing or processing, and displaying the data in the same manner as a hardwired commercial instrument. Designing a virtual instrument is faster and much more versatile than designing a hardware system.

This chapter was developed by National Instrument personnel: Erik Luther, Christopher Malato, and Vivek Nath.

CHAPTER 10

Control Design in LabVIEW

Laboratory Virtual Instrumentation Engineering Workbench (LabVIEW) provides the ability to design and implement dynamic system models with the Control Design and Simulation Module. This module contains Control Design functions that you can use to construct and analyze system models in the form of transfer functions, zero-pole-gain models, or state-space models. The module also contains Simulation capabilities, which are explained in the next chapter.

10.1 CONTROL DESIGN FUNCTIONS

The LabVIEW Control Design and Simulation Module provides virtual instruments (VIs) that you can use to create and develop control design applications. These VIs are found in LabVIEW by browsing to Control Design & Simulation ≫ Control Design in the Functions Palette. The palette includes VIs to construct and analyze mathematical models of dynamic systems.

Many of the functions provided in the Control Design VIs are also available as MathScript functions, which can be used in a MathScript Node to construct and analyze models.

10.2 CONTINUOUS VERSUS DISCRETE MODELS

Dynamic systems can be represented by either continuous or discrete models. A continuous model describes how a system behaves continuously with time, meaning the properties of a system can be obtained at any point in time. A discrete model describes how a system behaves at separate instants in time, meaning the properties of a system can only be obtained at those specific times.

Both continuous and discrete models can be created with the LabVIEW Control Design and Simulation Module. The exercises and examples in this chapter will focus only on continuous transfer function models, which have the following form:

$$H(s) = \frac{b_0 + b_1 s + \cdots + b_{m-1} s^{m-1} + b_m s^m}{a_0 + a_1 s + \cdots + a_{n-1} s^{n-1} + a_n s^n}$$

10.3   MODEL CONSTRUCTION
10.3.1 Constructing a Transfer Function Graphically
Control design models can be created graphically in LabVIEW with the VIs inside the Model Construction palette (Control Design & Simulation ≫ Control Design ≫ Model Construction in the Functions Palette).

To create a transfer function model, drop the CD Construct Transfer Function Model VI on the block diagram of a VI (Figure 10.1). The VI icon has a polymorphic selector below it that allows you to select the mode of operation. In this lesson, the Single-Input Single-Output (SISO) mode will be used.

The "Sampling Time (s)" input determines whether the transfer function model will be continuous or discrete. If this input is left unwired or has a value of zero wired to it, then the model will be continuous. If a positive nonzero value is wired, then the model will be discrete. Once a discrete model is created, the sampling time can be obtained with the CD Get Sampling Time from Model VI (Control Design & Simulation ≫ Control Design ≫ Model Information ≫ CD Get Sampling Time from Model.vi in the Functions Palette).

Example
Create the following transfer function graphically in LabVIEW (Figure 10.1).

$$H(s) = \frac{10s^2 + 7s}{0.5s^3 + 6s^2 + 7s + 1}$$

1. Place the CD Construct Transfer Function Model VI on a LabVIEW block diagram.
2. Create controls from the *Numerator* and *Denominator* terminals.
3. On the front panel, fill in the Numerator and Denominator controls with the polynomial coefficients of the transfer function $H(s)$. Coefficients are sorted in ascending order.
4. Connect the *Transfer Function Model* output of the VI to the CD Draw Transfer Function Equation VI to display the model equation on the front panel.

10.3.2 Constructing a Transfer Function with MathScript
Control design models can be created textually in LabVIEW with a MathScript Node. Models produced with MathScript are of the same data type as models produced graphically with the Control Design VIs. Models created from either of the two methods can be used interchangeably.

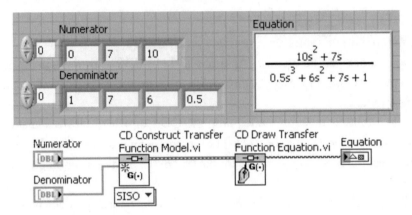

FIGURE 10.1: Creating a transfer function.

To create a transfer function model place a MathScript Node on the block diagram of a VI (Programming ≫ Structures ≫ MathScript Node in the Functions Palette). Use the *tf* command inside the node to construct the model.

Example
Create the transfer function model textually in LabVIEW with a MathScript Node

$$H(s) = \frac{10s^2 + 7s}{0.5s^3 + 6s^2 + 7s + 1}$$

1. Place a MathScript Node on a LabVIEW block diagram (Figure 10.2).
2. Type the following in the node (note that the coefficients are sorted in descending order in MathScript):
   - num = [10 7 0];
   - den = [0.5 6 7 1];
   - model = tf(num,den);
3. Create an output on the MathScript Node, and title it "model" to match the variable name defined in the node.
   a) Right-click the border of the node and select Add Output.
   b) To select the data type, right-click the new output and select Choose Data Type ≫ Addons ≫ TF Object.

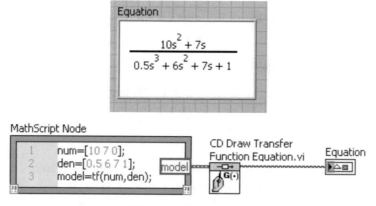

FIGURE 10.2: Creating a transfer function with MathScript.

4. Connect the *model* output of the node to the CD Draw Transfer Function Equation VI to display the model equation on the front panel.

10.4 MODEL INTERCONNECTION

Most control systems consist of multiple models that interact with each other. A system that is represented by a block diagram can be simplified by examining how the individual blocks are interconnected. The types of model interconnection that will be discussed here are Series, Parallel, and Feedback.

10.4.1 Series Interconnection

Two models, G1 and G2, are connected in series if the output of G1 is connected to the input of G2. The two models can be represented as a simplified model, G3 (Figure 10.3).

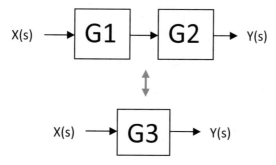

FIGURE 10.3: Series interconnection.

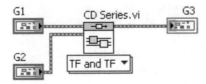

FIGURE 10.4: Connecting two models in series with LabVIEW.

You can interconnect models in series with the CD Series VI (Control Design & Simulation ≫ Control Design ≫ Model Interconnection ≫ CD Series.vi in the Functions Palette). The VI has two model inputs and one model output (Figure 10.4). The output value represents the series simplification of the two input models.

MathScript can be used to connect models in series with the *series* function:

$$G3 = \text{series} \ (G1, G2).$$

10.4.2 Parallel Interconnection

Two models, G1 and G2, are connected in parallel if they share the same input and their outputs are summed together. The two models can be represented as a simplified model, G3 (Figure 10.5).

You can interconnect models in parallel with the CD Parallel VI (Control Design & Simulation ≫ Control Design ≫ Model Interconnection ≫ CD Parallel.vi in the Functions Palette), as shown in Figure 10.6.

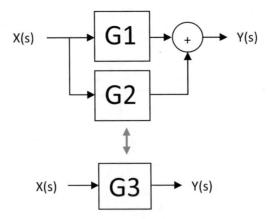

FIGURE 10.5: Parallel interconnection.

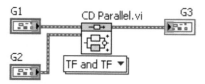

FIGURE 10.6: Connecting two models in parallel with LabVIEW.

MathScript can be used to connect models in parallel with the *parallel* function: G3 = parallel (G1,G2).

10.4.3 Feedback Interconnection

- Single model

    A single model, Gc, is in a closed-loop, or feedback, configuration if its output is connected back to its input (Figure 10.7). This closed-loop system can be represented as a simplified open-loop model, Go.

    You can simplify a closed-loop model to an open-loop model with the CD Feedback VI (Control Design & Simulation ≫ Control Design ≫ Model Interconnection ≫ CD Feedback.vi in the Functions Palette). Wire the Gc model to the "Model 1" Input, and leave the "Model 2" input unwired as shown in Figure 10.8.
- Two models

    Two models, G1 and G2, are in a closed-loop, or feedback, configuration if each model's output is connected to the other model's input. This closed-loop system can be represented as a simplified open-loop model, G3, as shown in Figure 10.9.

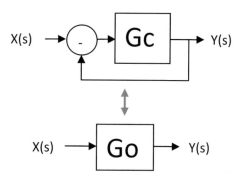

FIGURE 10.7: One model in a feedback loop.

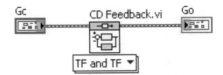

FIGURE 10.8: Creating a one-model feedback loop with LabVIEW.

You can simplify a two-model, closed-loop system to an open-loop model with the CD Feedback VI. Wire the G1 model to the "Model 1" input, and wire the G2 model to the "Model 2" input, as shown in Figure 10.10.

10.5   MODEL ANALYSIS
10.5.1 Time Response
Time response analysis is used to determine how a system behaves to certain inputs. You can analyze the time response to determine the stability of the system. The LabVIEW Control Design & Simulation module provides several VIs that can plot the time-domain output of a system or display parametric data about the behavior of the system.

10.5.2 CD Parametric Time Analysis
Use the CD Parametric Time Response VI (Control Design & Simulation ≫ Control Design ≫ Time Response ≫ CD Parametric Time Response.vi in the Functions Palette) to obtain

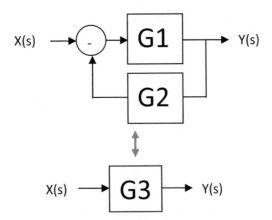

FIGURE 10.9: Two models in a feedback loop.

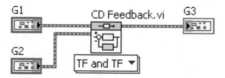

FIGURE 10.10: Creating a two-model feedback loop with LabVIEW.

numerical time response data for a system (Figure 10.11). You can wire a system model created with the model construction and model interconnection VIs to the input of this VI. The "Time Response Parametric Data" output of this VI contains the following data fields:

- *Rise time* (s): The time for the system response to rise from a lower threshold (10% default) to an upper threshold (90% default). Only valid with Step Response analysis.
- *Overshoot* (%): Percentage that the system response exceeds unity. Only valid with Step Response analysis.
- *Peak time* (s): The time required for the system response to reach the peak value of the first overshoot.
- *Steady-state gain*: The value around which the system response settles to a steady state. Only valid with Step Response analysis.
- *Settling time* (s): The time required for the system response to reach and stay within a threshold (1% default) of the final value. Only valid with Step and Impulse Response analysis.
- *Peak value*: The maximum value of the system response.

10.5.3 Analyzing a Step Response
The step response of a dynamic system measures how the output of the system responds to a step function input. A step function is defined as:

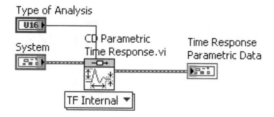

FIGURE 10.11: Parametric time response analysis of a model in LabVIEW.

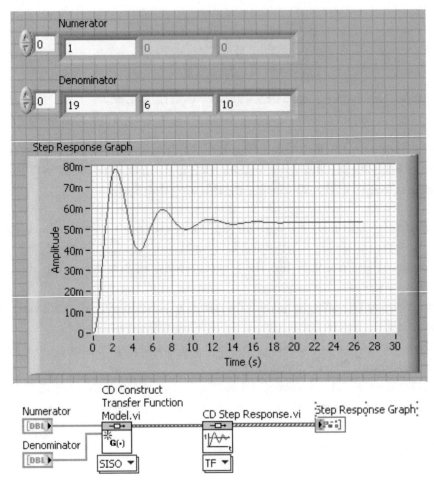

FIGURE 10.12: Step response of a transfer function.

$$u(t) = 0 \text{ when } t < 0$$
$$u(t) = 1 \text{ when } t \geq 0$$

You can use the CD Step Response VI (Control Design & Simulation ≫ Control Design ≫ Time Response ≫ CD Step Response.vi in the Functions Palette) to plot the time-domain step response of a system, as shown in Figure 10.12.

10.5.4 Analyzing an Impulse Response
The impulse response of a dynamic system measures how the output of the system responds to an impulse input signal.

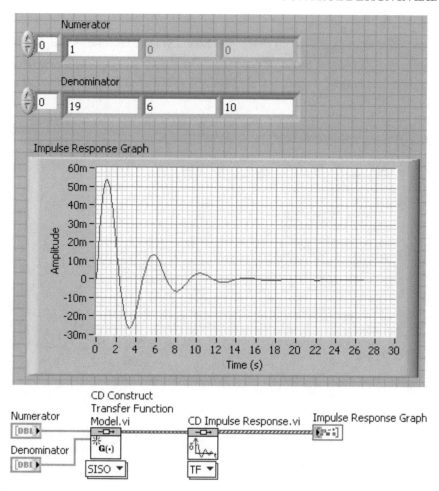

FIGURE 10.13: Impulse response of a transfer function.

1. For a continuous system, an impulse signal is defined as a unit-area signal with infinite amplitude and infinitely small duration. The signal value is zero at all other times.
2. For a discrete system, an impulse signal is defined as a pulse that has unit amplitude at the first sample period and zero amplitude at all other times.

You can use the CD Impulse Response VI (Control Design & Simulation ≫ Control Design ≫ Time Response ≫ CD Impulse Response vi in the Functions Palette) to plot the time-domain impulse response of a system, as shown in Figure 10.13.

Analyzing an Initial Response

• CD Initial Response

10.5.5 Frequency Response

Frequency response analysis is used to determine how a system responds to sinusoidal inputs of unit amplitude, zero phase, and varying frequencies. The frequency response of a system can reveal information about the system's behavior and stability across a wide frequency range. This is some-

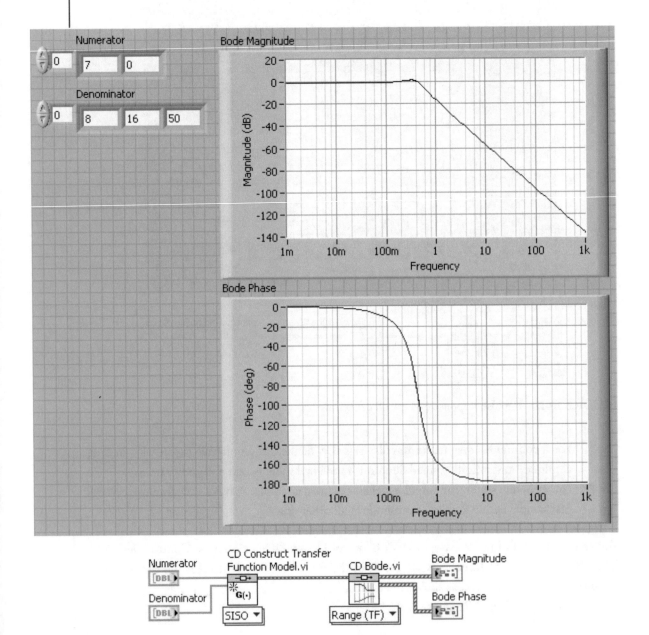

FIGURE 10.14: Bode magnitude and phase of a transfer function.

times more useful than time response analysis, which shows the behavior of a system under a single set of circumstances.

The LabVIEW Control Design and Simulation Module provides VIs that can analyze the Bode frequency, Nichols frequency, and Nyquist stability of a system.

You can use the CD Bode VI (Control Design & Simulation ≫ Control Design ≫ Frequency Response ≫ CD Bode.vi in the Functions Palette) to plot the frequency-domain magnitude and phase responses of a system. Wire the model to CD Bode.vi, then create indicators from the VI output terminals, as shown in Figure 10.14. Similarly, you can display Nichols plots or Nyquist plots for a system model by wiring the model to the appropriate VI, each found in the Frequency Response palette.

10.6   REVIEW EXERCISES

1. Create a LabVIEW VI that constructs the three continuous transfer function models, $A(s)$, $B(s)$, and $C(s)$:

$$A(s) = \frac{1}{m_1 s^2 + bs + k_1 + k_2}$$

$$B(s) = \frac{k_2}{m_2 s^2 + k_2}$$

$$C(s) = k_2$$

where

$m_1 = 10$
$m_2 = 4$
$k_1 = 12$
$k_2 = 7$
$b = 6$

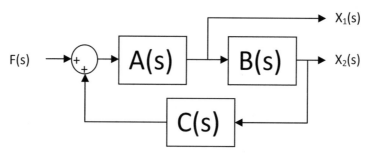

FIGURE 10.15: Feedback system.

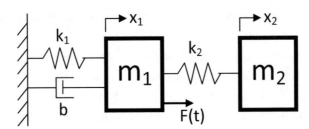

FIGURE 10.16: Duel spring-mass system.

Use CD Construct Transfer Function Model.vi with the "SISO (Symbolic)" mode selected. Use CD Draw Transfer Function.vi to display the equation for each model.

2. Consider the following block diagram system representation that contains the models $A(s)$, $B(s)$, and $C(s)$ from Exercise 1 (see Figure 10.15):

3.

   a) Use model interconnection to create the transfer function $X_1(s)/F(s)$, using the three transfer function models created in Exercise 1.

   b) Use model interconnection to create the transfer function $X_2(s)/F(s)$.

4. The figure in Exercise 2 represents the spring-mass-damper system below (Figure 10.16):

   Use CD Parametric Time Response.vi to plot the response, $x_1(t)$, and to calculate the settling time of the system.

This chapter was developed by National Instrument personnel: Erik Luther, Christopher Malato, and Vivek Nath.

CHAPTER 11

Simulation in LabVIEW

As discussed earlier, a model of a control system can be developed using the Laboratory Virtual Instrumentation Engineering Workbench (LabVIEW) Control Design and Simulation Module. It is extremely useful to be able to recreate and analyze the behavior of a dynamic system in software before implementing a model in hardware. This makes simulation an important part of a control system. It offers easy reusability and the ability to observe the behavior of a dynamic system.

Control design can be used to identify and design the model of a control system. Simulation can then be used to validate the system under real-world constraints. It can be used to model linear, nonlinear, discrete, and continuous control systems in LabVIEW.

11.1 SIMULATION LOOP
The first step in performing simulation in LabVIEW is creating a *simulation loop*. It brings together differential equation solvers and timing features into a structure similar to a while loop. All simulation functions must be placed in the loop on a LabVIEW block diagram (Figure 11.1).

The loop consists of three main parts:

1. Input node
   This node allows simulation parameters to be programmatically configured.

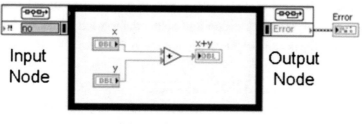

FIGURE 11.1: A simulation loop.

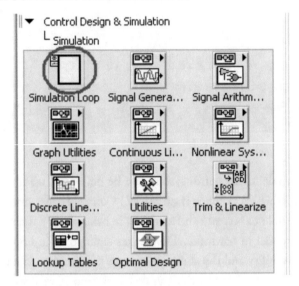

FIGURE 11.2: The simulation loop is found in the "Simulation" palette.

2. Main loop

   This is where the system to be simulated is placed.

3. Output node

   This node returns any errors that may have occurred in the loop. If an error occurs while the simulation loop is running, the simulation stops and returns the error in this output.

11.2   CREATING A SIMULATION LOOP

Follow these steps in to create a simulation loop.

1. You can find the simulation loop in the functions palette under Control Design and Simulation ≫ Simulation (Figure 11.2). When selected, the mouse cursor becomes a special pointer that can be used to enclose the section of code that needs to be repeated.

2. Left-click once on the block diagram to define the top-left corner of the loop. Left-click again to define the bottom-right corner and create the simulation loop (Figure 11.3).

11.3   CONFIGURING A SIMULATION

There are many parameters that can be configured for a given simulation loop. You can view a configuration window for these parameters by double-clicking on the input node. The two types of parameters that can be configured here are simulation parameters and timing parameters.

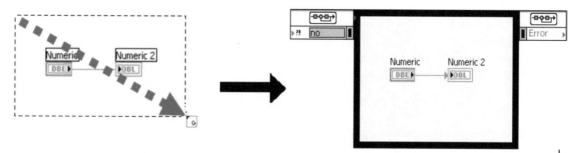

FIGURE 11.3: Creating a simulation loop.

11.4   SIMULATION PARAMETERS TAB

The different simulation parameters (Figure 11.4) that can be configured are:

- Simulation time

  Specifies for what period of "simulation time" how long the simulation should run. This time does not necessarily dictate the computation time of the simulation.

  Tip: Inserting 'inf' as the Final Time will let the loop run indefinitely until halted by the user.

- Solver method

  Specifies what ordinary differential equation (ODE) solver is used to solve dynamic system models in the simulation. A wide variety of solvers are available. High-order ODE solvers, such as Runge-Kutta 4, usually are more accurate than low-order ODE solvers, such as Euler. In general, you can use fewer time steps and larger step sizes with a high-order ODE solver to obtain the accuracy you need. Using fewer time steps decreases the effects of round-off in the solution and potentially reduces the amount of time needed to compute the solution.

- Time step and tolerance

  These settings control the window of time steps used by LabVIEW. Typically, the default settings will suffice but adjust them if necessary.

- Discrete time

  Although the Default Auto Discrete Time option will typically work for most simulations, you can force LabVIEW to use a specific step size here.

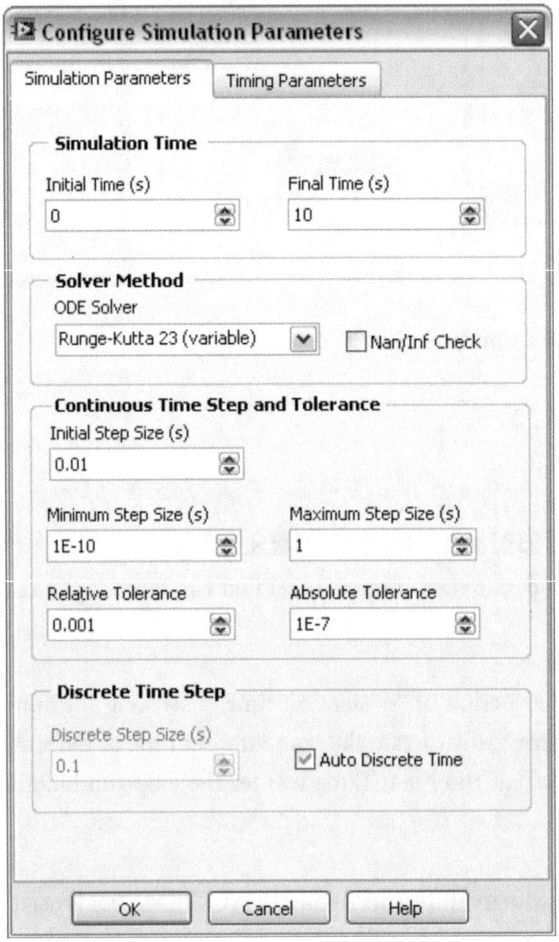

FIGURE 11.4: Simulation parameters dialog.

11.5   TIMING PARAMETERS TAB

Figure 11.5 shows the Simulation Timing Parameter dialog window.

- Synchronized timing
  The loop timing can be synchronized to a timing source if required. If this box is checked, the loop will solve the equation as fast as the CPU can.
- Loop timing source
  When synchronizing the loop to a timing source, the source type can be selected here.
- Loop timing attributes

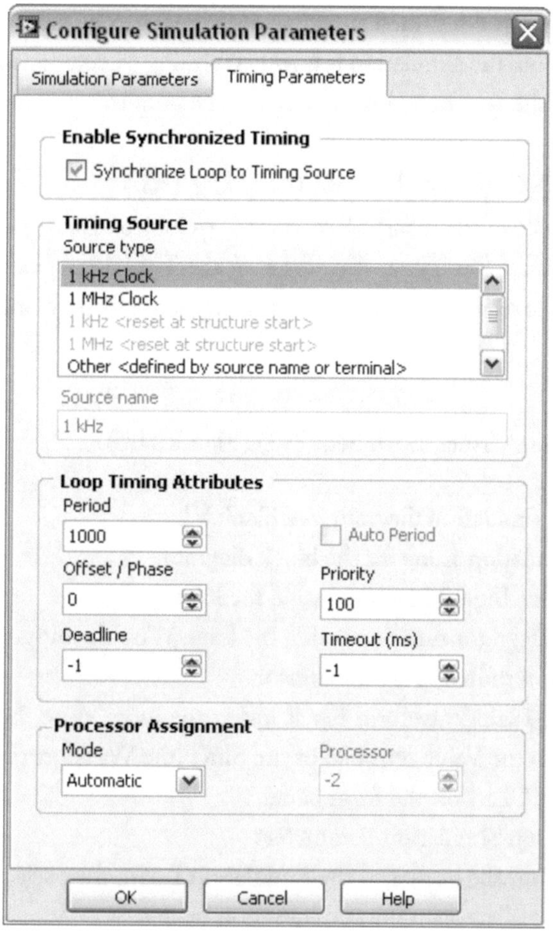

FIGURE 11.5: Timing dialog.

These options control how the loop executes with respect to the selected timing source. The value for the period specifies the simulation loop rate with respect to the timing source.

11.6   GENERATING SIMULATION SIGNALS

Providing a simulated signal to a control system is useful for analysis of its behavior based on its response. The Signal Generation palette (Control Design & Simulation ≫ Simulation ≫ Signal Generation in the Functions palette) contains several functions that generate periodic signals inside a simulation loop.

Once the simulation is confirmed, a real-world signal can be substituted for the simulation signal. In this configuration, the simulation is now performing calculations based on actual data and is very useful for testing the system before controlling the output.

11.7   DISPLAYING SIMULATION OUTPUT

The Simulation Time Waveform chart (Functions ≫ Control Design & Simulation ≫ Simulation ≫ Graph Utilities ≫ SimTime Waveform in the Functions palette) is a special numeric indicator that displays one or more plots. Waveform charts can display single or multiple plots.

Example

Create a simple Simulation virtual instrument (VI) (Figure 11.6).

1. Create a simple simulation diagram in a blank VI:
   a) Place a Simulation Loop on the block diagram.
   b) Place the Sine Signal function inside the loop.
   c) Place the Integrator function inside the loop. Wire the output terminal of Sine Signal to the input terminal of the Integrator.
   d) Place a SimTime Waveform block inside the loop. Wire the output terminal of the Integrator to the Value terminal of the SimTime Waveform block. This action creates a Waveform Chart on the front panel.
2. Configure the loop Simulation Parameters
   a) Right-click on the border of the Simulation Loop, then select "Configure Simulation Parameters . . ." to open the configuration dialog.
   b) Set the Simulation Parameters according to the following values:
      • ODE Solver = "Runge-Kutta 1 (Euler)"

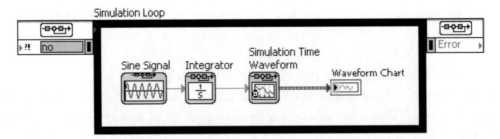

FIGURE 11.6: A simulation loop that integrates a sine wave.

- Step Size (s) = "1"
- Leave all other parameters with the default values.

   c) Click "OK" to close the configuration dialog.

   d) Run the VI, and observe the results in the Waveform Chart on the front panel. Note that the values are very inaccurate. This is because of the large step size used.

3. Improve the loop Simulation Parameters

   a) Reopen the configuration dialog, and change the value of Step Size (s) to "0.25."

   b) Run the VI, and observe the improved results. Greater reduction in step size, as well as using a more sophisticated ODE Solver, will improve the results further.

4. Synchronize the Timing Parameters with the Simulation Parameters.

   a) In the configuration dialog set the value of Step Size (s) to "0.01" on the Simulation Parameters tab.

   b) On the Timing Parameters tab, set parameters according to the following values:

- Synchronize Loop to Timing Source: check this box
- Timing Source = "1 kHz Clock"
- Period = 10
- Leave all other parameters with the default values. This will make the loop run one iteration (each 10 ms), which is synchronized with the 10-ms step size specified in the Simulation Parameter.

   c) Close the dialog, then run the VI. The waveform chart will now update in "real time."

11.8   IMPLEMENTING TRANSFER FUNCTIONS

Transfer functions can be implemented in a Simulation loop with the Transfer Function block (Functions ≫ Control Design & Simulation ≫ Simulation ≫ Continuous Linear Systems ≫ Transfer Function in the Functions Palette). Create and configure a transfer function by following these steps:

1. Place a Transfer Function block in a Simulation loop. Double-click the block to open the configuration dialog.
2. Select "Transfer Function" in the Parameter Name list. Select "Configuration Dialog Box" for Parameter source.
3. Enter values for the Numerator and Denominator.
4. Click "OK" to exit the dialog. The transfer function will now implement the model selected in the dialog.
5. Wire signals from other parts of the diagram to the input and output of the transfer function.

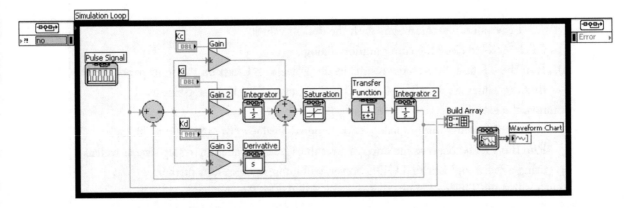

FIGURE 11.7: Example of a control system in LabVIEW.

In control design, models such as transfer functions are represented by wires on the block diagram. On the other hand, in the simulation loop, models are represented by blocks. Simulation provides models for both continuous and discrete systems as well as nonlinear systems.

Other functions that can be implemented in the simulation loop include:

- State-space
- Zero-pole-gain
- Integrator
- Derivative
- Transport Delay

Figure 11.7 presents an example of running a simulation of a feedback control system in LabVIEW.

This chapter was developed by National Instrument personnel: Erik Luther, Christopher Malato, and Vivek Nath.

CHAPTER 12

LabVIEW Control Design and Simulation Exercise

LabVIEW, which stands for Laboratory Virtual Instrumentation Engineering Workbench, is a graphical computing environment for instrumentation, system design, and signal processing.

The Control Design and Simulation (CDSim) module for LabVIEW can be used to simulate dynamic systems. To facilitate model definition, CDSim adds functions to the LabVIEW environment that resemble those found in SIMULINK. There is also the ability to use m-file syntax directly in LabVIEW through the new MathScript node.

The purpose of this tutorial is to introduce you to LabVIEW and give you experience in simulating dynamic systems. In the first section, you will build a model of the open-loop system for the second-order plus time delay process $G(s) = \dfrac{2e^{-s}}{(10s + 1)(5s + 1)}$ and determine the unit set point and unit disturbance responses. In the second section, you will build a closed-loop model of the same process. After the closed-loop model is constructed, you should simulate the unit disturbance response and the unit setpoint response for two different proportional, integral, and derivative (PID) controller tuning methods, ITAE (setpoint) and ITAE (load) [1].

Log onto one of the computers. Click Start ≫ National Instruments ≫ LabVIEW. Open LabVIEW.

To start a new program (called VI for Virtual Instrument), click "Blank VI" as shown in Figure 12.1. Two LabVIEW windows, the Front panel window and the Block diagram window, will appear as shown in Figure 12.2.

Click in the Block Diagram to view the area where graphical programs are written. Right-click inside the Block Diagram to view the palette of Functions used in creating programs (Figure 12.3). Select the Control Design & Simulation ≫ Simulation palette to view the library of simulation functions.

In the next section, you will build a model of the open-loop system for the process mentioned earlier, and determine the unit setpoint and unit load responses. The following steps will guide you through the discussed tasks.

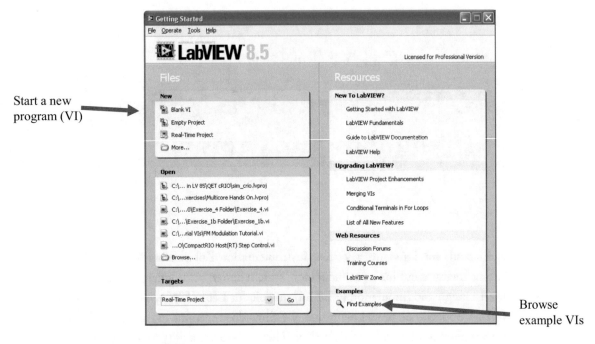

Start a new program (VI)

Browse example VIs

FIGURE 12.1: Initial LabVIEW Screen.

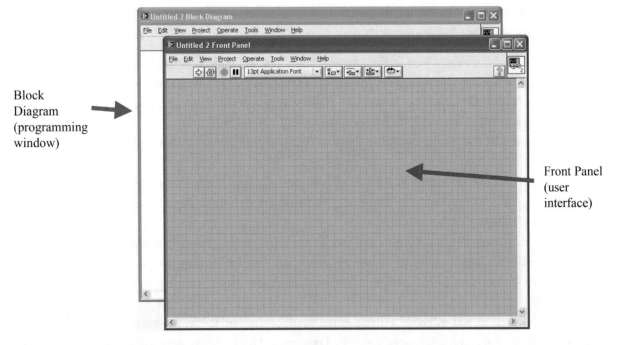

Block Diagram (programming window)

Front Panel (user interface)

FIGURE 12.2: LabVIEW new VI.

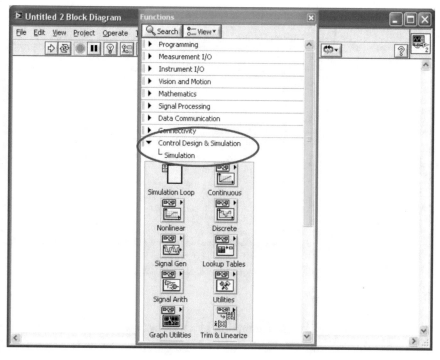

FIGURE 12.3: Simulation functions palette.

12.1   CONSTRUCTION OF AN OPEN-LOOP BLOCK DIAGRAM

1. Open a new VI by selecting File ≫ NewVI. The new window will be titled *untitled*. You will build your closed-loop model in the block diagram. Save the empty model by choosing File ≫ Save. Name the model *examplesim*. From this point on, the model will be referred to as *examplesim*.

2. Click on the block diagram, then right-click to bring up the Functions palette. From the Simulation subpalette, click-and-drag a Simulation loop on the block diagram (Figure 12.4).

3. Place the Transfer Function and Transport Delay blocks from the "Continuous" palette, respectively, to *Examplesim*. Connect the output of the Transfer Function block to the input of the Transport Delay block. Click on the "Transfer Function" label and rename to "Process TF." This block represents the process. Note that, in this problem, the process is $G(s) = G_v G_p G_m$, not G_p.

   Open the dialog box of Process TF by double-clicking on it (Figure 12.5). Specify Numerator as [2] and Denominator as [1 15 50], which defines the transfer function as

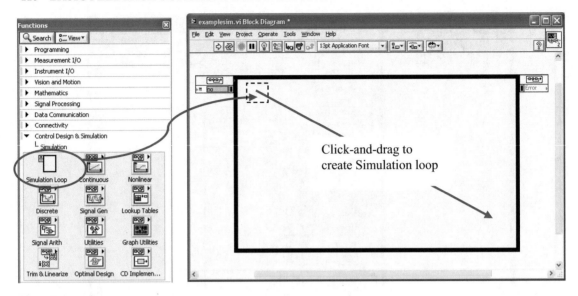

FIGURE 12.4: Creating a simulation loop.

$2/(50s^2 + 15s + 1)$. The Transfer Function block allows specification of vectors for the numerator and the denominator from either a Configuration Dialog Box or a Terminal from the block diagram. The vector elements are treated as the coefficients of *ascending* powers of s in the polynomials representing the numerator and denominator of the transfer function. To see the denominator polynomial of s completely displayed in the block's icon, you may

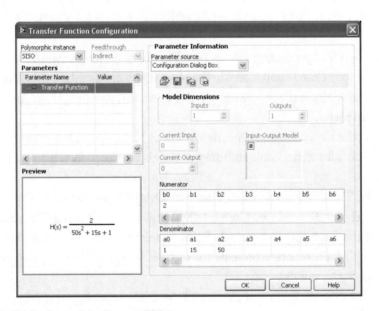

FIGURE 12.5: Dialog box of the Process TF.

have to resize the block's icon. Double-click on the Transport Delay and set Time delay to unity (one). Note that the Transport Delay block can be used to represent other types, such as measurement delay.

4. Copy the Process TF and Transport Delay blocks and place the copies slightly above the originals. The copies will automatically change names to "Process TF1" and "Transfer Delay1." To quickly copy the original blocks, select both of them, hold the CTRL key, and drag using the left mouse button. Rename the Process TF1 block "Load TF." In this example, G_p and G_L are the same, so the Numerator and Denominator parameters in the dialog box of Load TF are not changed (Figure 12.7).

5. Place a copy of the Summation block, located in the "Signal Arithmetic" palette, to the right of the Transport Delay block. Right-click on the summation block and select Visible Items ≫ Label to see the label "Summation." Connect the output from each Transport

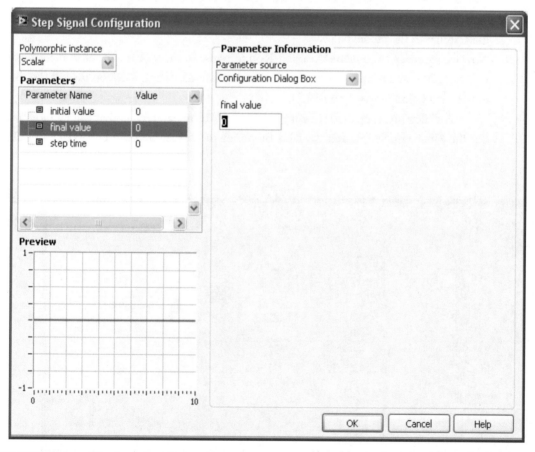

FIGURE 12.6: The Step Signal block generates a step function with specified initial value, final value, and step time.

Delay block to the input of the Summation block. The number of inputs and their polarity can be modified from the dialog box. Later in the tutorial, you will be required to do this.

6. Place a SimTime Waveform graph, from the "Graph Utilities" palette, to the right of the Summation block. Connect the output of the Summation block to the input of the Sim-Time Waveform block.

7. Place a Step Signal block, from the "Signal Generation" palette, to the left of Load TF and connect it to the input of Load TF. The Step Signal block generates a step function. The initial value, final value, and step time (time at which the step occurs) of the function can be specified (Figure 12.6). For now, double-click to open its dialog box and set the initial value, final value, and step time to zero, i.e., disabling the block. Rename the block "L."

8. Place a copy of L to the far left of Process TF and rename the new block "U." Connect U to the input of Process TF. Double-click on U and set Step time to 0, Initial value to 0, and Final value 1. U will generate a unit step function in the manipulated variable at time zero. The model developed to this point is a model of the open-loop system. It should look similar to the model below.

9. Now we are ready to simulate the open-loop response of the system. To select the integration technique and parameters for use during simulations, double-click on the left terminal of the Simulation loop (Figure 12.7).

A dialog box is opened showing all the simulation parameters that can be modified. Set the Final Time to 50 and the Max Step Size to 1. Note that the LabVIEW Simula-

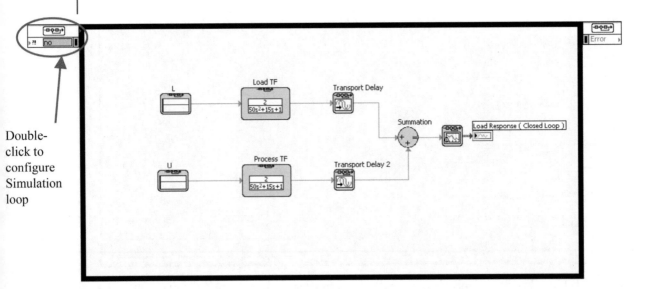

Double-click to configure Simulation loop

FIGURE 12.7: Open-loop block diagram.

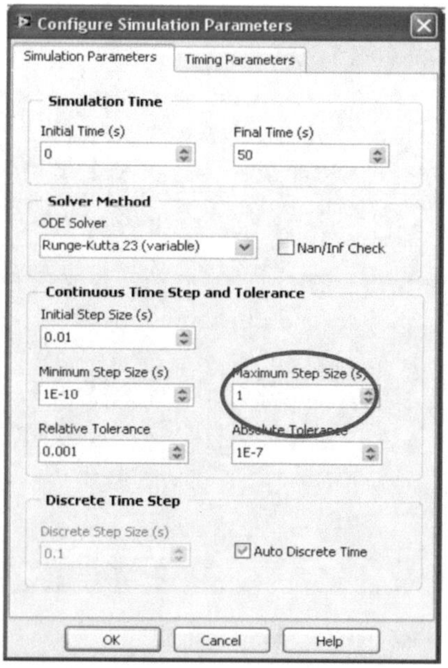

FIGURE 12.8: Simulation Parameter setting Final Time and Max Step Size.

tion loop includes an ordinary differential equation (ODE) solver. The maximum step size determines the largest step LabVIEW uses in numerically integrating the ODE. Because this system is easy to numerically integrate, a Max Step Size of 1 will result in a smooth curve (Figure 12.8). Larger step sizes will produce more jagged curves.

Run the simulation by clicking the *Run* arrow on either the Front Panel or the Block Diagram (Figure 12.9). Hint: Ctrl E switches between the Front Panel and Block Diagram.

10. The response of the open-loop system for a unit step input will be automatically plotted, as shown in Figure 12.10. Double-click on the title to change the name of the plot. You can also right-click on the plot to view axis settings, autoscaling, and other plot parameters.

FIGURE 12.9: Run simulation by clicking the *Run* arrow.

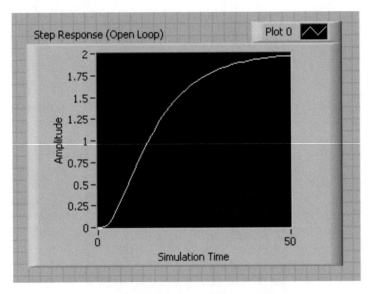

FIGURE 12.10: Unit step response of open-loop system.

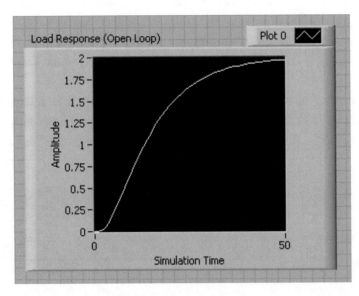

FIGURE 12.11: Unit load response of open-loop system.

11. Now simulate the open-loop unit load response. Double-click on U and set Final value to 0. This eliminates the unit step in the manipulated variable. Double-click on L, and set Step time to 0 and Final value to 1. This creates a unit load step. Again, hit the *Run* arrow to begin the simulation. The Front Panel will show the response. Double-click on the graph title to replace Step Response (Open Loop) with Load Response (Open Loop). Figure 12.11 shows the resulting plot. Note that the open-loop setpoint response and load response are the same. Why is this expected?

12. If you are not continuing to the next section, save the file *examplesim* so that you can use it in constructing a closed-loop block diagram.

12.2   CONSTRUCTION OF CLOSED-LOOP BLOCK DIAGRAM

1. Open the file *examplesim* if it is not already open.
2. Click on the connection between the U block and Process TF block, and then delete it. Rename the U block to "R." This block will be used to produce a step change in the setpoint.
3. Place a copy of the Sum block (Figure 12.12) to the right of R. It will automatically be given the label "Summation2." Open its dialog box and change the lower input from +

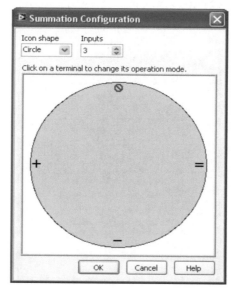

FIGURE 12.12: Summation configuration.

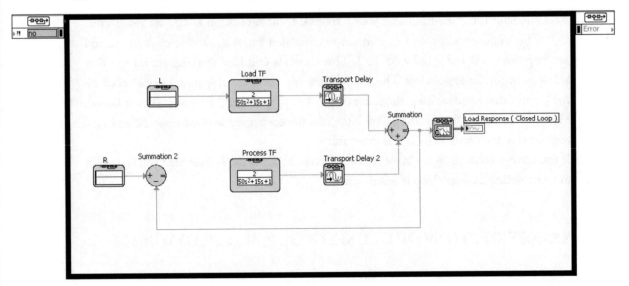

FIGURE 12.13: Partially completed closed-loop diagram.

to −. The top input will have a + located to the right of it, whereas the bottom input will have a − located above it. Therefore, the output of Summation2 will be the top input minus the bottom input.

Connect the output of R to the top input of Sum1. Also, connect the output from Sum to the bottom input of Sum1. This can be done by clicking on the bottom input of Sum1

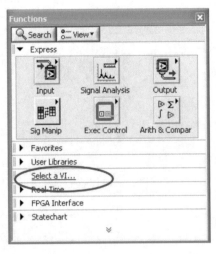

FIGURE 12.14: From the Functions palette, click on "Select a VI."

and dragging the arrow to the output of Sum. The output of Sum1 is the error between the setpoint, R, and the controlled variable, C. Your model should look like Figure 12.13.

4. Now right-click in the Block Diagram to bring up the Functions palette, and click on "Select a VI" as shown in Figure 12.14.

This action allows you to bring in any user-defined LabVIEW VI into your current program. Click on the path, then select the "PID Controller" and drag it to the right of the newest sum block. Connect the output of Summation2 to the input of PID controller and the output of PID controller to the input of Process TF. Double-click on the PID controller and enter the ITAE (load) controller settings given in Table 12.1. Please note that PID controller settings are K_c, τ_i, and τ_D, where $P = K_c$, $I = K_c/\tau_i$, and $D = K_c\tau_D$, so numerical values of P, I, and D should reflect these definitions. The model you have developed represents the closed-loop system. Your model should now look similar to Figure 12.15.

Note that text has been added to the block diagram shown on the previous page. Simply by double-clicking on a point in *examplesim* and typing, text is added to the diagram at the point where you clicked. The E, P, X1, X2 text in the block diagram have no effect on its operation.

5. An important feature of LabVIEW is interactivity. We can use this capability to make the PID controller gains interactive from the Front Panel, rather than having to edit them on the Block Diagram. Double-click on the PID Controller.vi and (Figure 12.16) change Parameter Source from *Configuration Dialog Box* to *Terminal*.

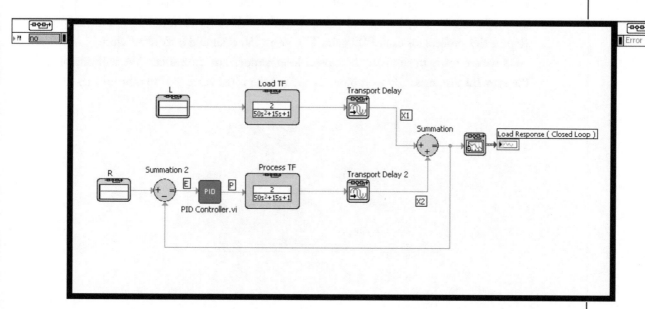

FIGURE 12.15: Closed-loop diagram.

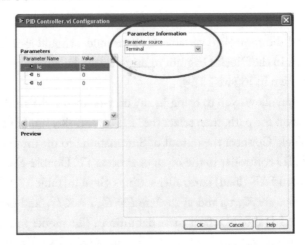

FIGURE 12.16: Controller PID configuration VI.

6. Use Ctrl H to bring up Help on the PID Controller.vi and see where the gain terminals are located (they come in from the top). Right-click on each terminal and select Create >> Control to automatically wire a control to the terminal. The inputs to the PID Controller. vi should look like Figure 12.17.

   By default, LabVIEW creates a standard Numeric control, but it can easily be changed. Go to the Front Panel, right-click on the control, select Replace >> Horizontal Pointer Slide. Then right-click on the new control and Visible Items >> Digital Display. In this manner, gains may be entered either from the slide or typed in to the numeric control. Repeat this process for each PID gain. The Front Panel should look like Figure 12.18.

7. Now we are ready to simulate the closed-loop response of the system. We will start with the setpoint response. Click on block L and set the Final value to 0 so that no step in the

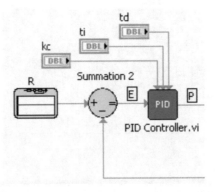

FIGURE 12.17: Inputs to the PID Controller.vi.

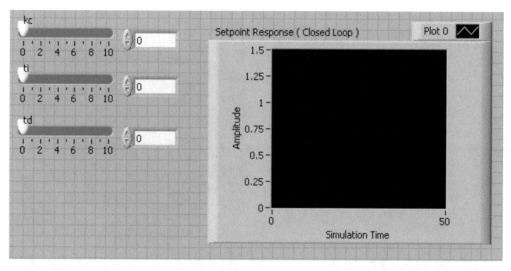

FIGURE 12.18: The Front Panel view of the example.

load will occur. Create a step in the setpoint by clicking on R and setting the Final value to 1. Run the simulation. The resulting graph (Figure 12.19) will be for the unit setpoint response, because L (disturbance) has been disabled.

8.   Now simulate the unit load response. Double-click on R and set Final value to 0. Double-click on L and set Final value to 1. Run the simulation again. Label this plot in the same manner as the previous one, except that you need to replace the title Setpoint Response with Load Response. Figure 12.20 shows the resulting plot.

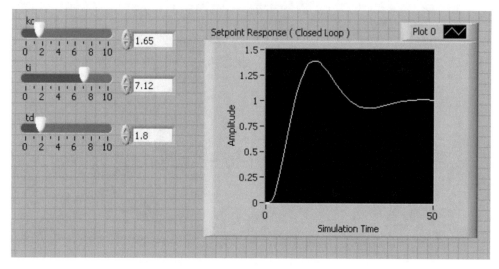

FIGURE 12.19: Unit setpoint response for ITAE (load) settings.

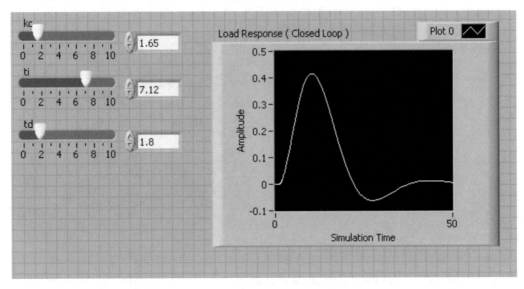

FIGURE 12.20: Unit load response for ITAE (load).

9. To save data for plotting, examine the changes made to the VI, saved as *examplesim with file.vi*, as in Figure 12.21.
10. Different settings may compare the two data sets to each other, using *Read and Plot Data.vi* as shown in Figures 12.22 and 12.23.

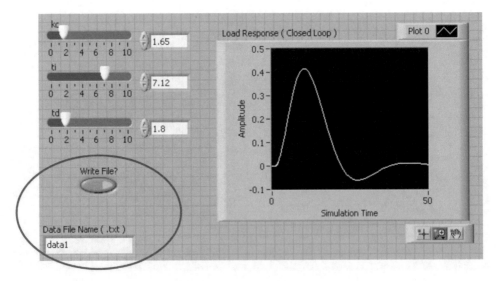

FIGURE 12.21: Saving data for plotting.

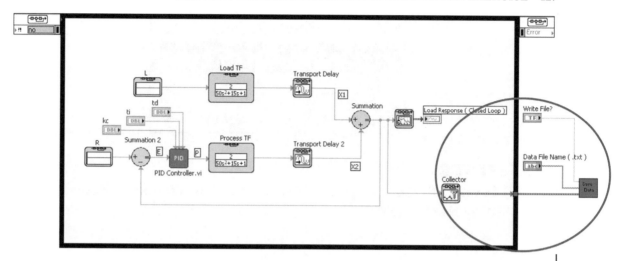

FIGURE 12.22: To compare two data sets to each other

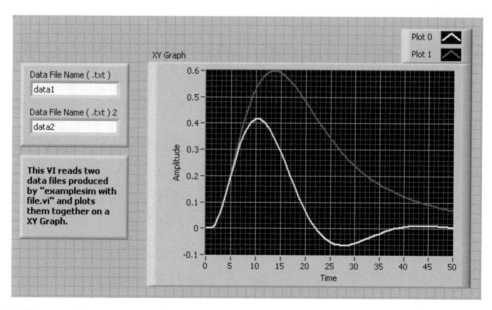

FIGURE 12.23: Comparison of PID controllers with ITAE settings of the two data sets.

TABLE 12.1: Gains for closed-loop exercise			
	K_C	t_I	t_D
ITAE (load)	1.65	7.12	1.80
ITAE (set point)	1.07	16.25	1.55

REFERENCE

[1] Seborg, D. E., Edgar, T. F., and Mellichamp, D. A., *Process Dynamics and Control*, 2nd ed., Wiley, New York (2004).

LABVIEW CONTROLS TUTORIALS

[1] http://zone.ni.com/devzone/cda/tut/p/id/6368, based on Prof. Dawn Tilbury's tutorials from the University of Michigan.

[2] http://cnx.org/content/col10401/latest/ NI LabVIEW training course on Rice's Connexions site.

[3] http://techteach.no/labview/ by Finn Haugen.

This chapter was developed by National Instrument personnel: Erik Luther, Christopher Malato, and Vivek Nath.

· · · ·

CHAPTER 13

Cardiac Control

Cardiac control is a complex, involved process. To model its control, one must consider input from the neural centers, the humoral controls, resistance of the entire body, and numerous feedback loops to various parts of the central nervous system (CNS) [1]. Blood pressure is considered for the derivation the transfer functions discussed in this paper, because of the applicability of the resulting transfer function. Parameters for various pressures have already been determined based on empirically gathered data.

The cardiovascular system is composed of the heart, the blood vessels (vasculature), and the plasma of the blood with red and white blood cells. The role of the cardiac system is to pump the blood, containing oxygen, nutrients, immune cells, and other regulatory molecules, to the tissues and organs of the body.

The human heart consists of four chambers, two atria and two ventricles, arranged in two, half functions (left and right heart) that serve as independent pumps. The right atrium receives blood from the venae cavae and sends it to the right ventricle from where the blood is pumped to the lungs for oxygenation. The left atrium of the heart receives oxygenated blood from the lungs where it is moved to the left atrium, which pumps the oxygenated blood to the tissues throughout the body. The heart valves and veins ensure unidirectional blood flow [4].

13.1 CARDIAC PARAMETERS
13.1.1 Heart Rate
The sinoatrial node (S-A) intrinsically sets the pace of the heartbeat. The signal for contraction starts when the SA node fires an action potential that spreads, through intermodal pathways, to the atrioventricular (A-V) node, the A-V bundle, the bundle branches, and the Purkinje fibers. The pace set by the S-A node is 188 beats/min ± tolerances. The parasympathetic and sympathetic branches (vagal nerves) of the autonomic nervous system influence the heart rate through antagonistic control. The parasympathetic stimulation serves to decrease the heart rate, whereas the sympathetic stimulation causes an increase. To achieve a resting heart rate of 70 beats/min, tonic parasympathetic activity must slow down the intrinsic rate of the S-A node.

The heart rate is also influenced by the respiratory cycle. Respiratory sinus arrhythmia refers to the normal slowing down of heart rate during expiration (breathing out) and speeding up of heart rate during inspiration. Heart rate is normally controlled by centers in the medulla oblongata. One of these centers, the nucleus ambiguous, increases parasympathetic nervous system input to the heart via the vagus nerve. The vagus nerve decreases heart rate by decreasing the rate of S-A node firing. Upon expiration, the cells in the nucleus ambiguous are activated and heart rate is slowed down. In contrast, inspiration triggers inhibitory signals to the nucleus accumbens and consequently the vagus nerve is not stimulated.

Atrial pressure is also a factor influencing heart rate. An increase in atrial pressure causes an increase in heart rate, because the atrial stretch receptors, which are part of the "Bainbridge reflex," transmit their afferent signals to the CNS through the vagus nerve. Other secondary factors influencing the heart rate are the partial pressures of carbon dioxide (pCO_2) and oxygen (pO_2). Peripheral chemoreceptors, located in the carotid and aortic arteries, respond to radical changes in arterial pO_2. A dramatic fall in pO_2 would stimulate ventilation, although this happens only during unusual physiological conditions, such as ascending to high altitude. Central chemoreceptors of the medullary respiratory control system, on the other hand, are highly responsive to pCO_2. When arterial pCO_2 increases, CO_2 crosses the blood-brain barrier quite rapidly and activate central chemoreceptors. These receptors then signal the central pattern generator to increase the rate and depth of ventilation.

13.1.2 Stroke Volume

Stroke volume (SV) is the volume of blood pumped by one ventricle in one contraction. Stroke volume may be measured and calculated by taking the volume of blood before contraction and subtracting the volume of blood after contraction. Alternatively, stroke volume (SV) is the difference between end diastolic volume (EDV) and end systolic volume (ESD), as shown by the following equation.

$$SV = EDV - ESD$$

Stroke volume relates to the difference between filling and emptying of the heart. The ventricular filling pressure and the amount of ventricular compliance determine the amount of filling. The emptying of the heart relates the ability of the heart muscle to contract, develop tension, and shorten. EDV is normally determined by venous return: the compression or contraction of veins returning blood to the heart. The ventricular afterload is controlled by arterial pressure and ventricular elastic volume, and affects the amount of emptying of the heart.

13.1.3 Cardiac Output

Cardiac output (CO) is the product of stroke volume and heart rate, of which heart rate is the more influential factor. Therefore, those factors that influence the heart rate also have a secondary control mechanism by increasing cardiac output. Cardiac output is the volume of blood pumped per ventricle per unit time. Cardiac output may also be viewed as being under secondary control of the autonomic nervous system.

$$CO = (SV)(HR)$$

$$MCAP = (CO)(PR)$$

where MCAP is the mean central arterial pressure and PR is peripheral resistance.

13.1.4 Contractility

Contractility is a major factor in determining the performance of the heart. Contractility is affected through the isotropic effect, whether positive (epinephrine) or negative. Although sympathetic stimulation causes an increase in cardiac contractility, changes in systolic pressure with vagal and parasympathetic efferent stimulations also affect contractility. These effects ultimately affect the stroke volume of the heart [5].

13.1.5 Preload and Afterload

The degree of myocardial stretch before contraction is called the preload on the heart because this stretch represents the load placed on cardiac muscles before they contract. During ventricular systole, ventricular force must overcome the myocardial stretch and the resistance created by the blood filling the arterial system. The combined load of EDV and arterial resistance during ventricular contraction is known as afterload. An increased in afterload can occur in elevated arterial blood pressure and loss of the ability of the aorta to stretch (compliance). To maintain constant stroke volume when afterload increases, the ventricle must increase its force of contraction, leading to an overall increase in energy adenosine-5'-triphosphate demand [4].

13.1.6 Autonomic Control

The heart rate is controlled by the autonomic nervous system, which is composed of the sympathetic and parasympathetic systems. The sympathetic system is mainly responsible for the high rate of enervation of the heart rate. Its signal is passed via the ganglions in the pre- and postganglionic fibers. The parasympathetic system is responsible for the reduction of enervation of the heart; if the signal from the vagus nerve were to be blocked, the heart rate would then increase to approximately 190 beats/min.

13.2   CARDIAC CONTROL DIAGRAM

Figure 13.1 shows the block diagram of the general cardiac control system. The heart has been split into four blocks: the right and left heart, and the SA and AV nodes. Doing so allows for better modeling of heart performance when coupled to the pulmonary system. In addition, control of the SA node can be considered an area of study—thus, modeling it as a block in the overall diagram provides utility in its design. The systemic circulation has been modeled as one block in regards to the heart. Later, it will be seen that the systemic resistance has been modeled as a four-element Windkessel. Feedback from the lungs factors into the heart rate in the form of arterial pressure and signals from stretch receptors located in the chest. Additional feedback loops into the CNS take the form of baroreceptors and chemoreceptors feeding into the medulla. Vasomotor tone from the systemic resistance also feeds into the medulla.

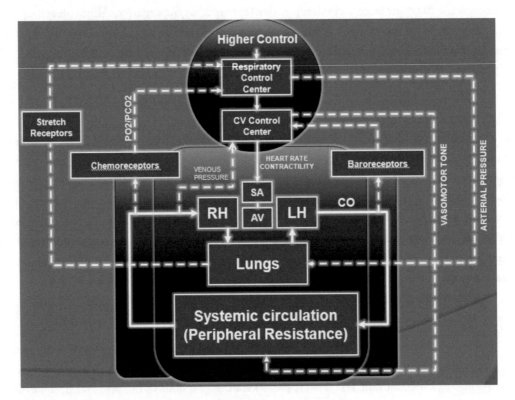

FIGURE 13.1: Block diagram of the general cardiac control system.

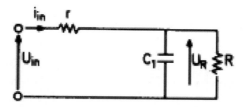

FIGURE 13.2: Three-element arterial bed.

Let us review the cardiovascular control of the left heart in terms of pressure. Pressure was chosen to be the focus of the transfer equations discussed below because of the relative ease at which pressure values may be empirically determined for the cardiovascular system under various states variable. The cardiovascular system considered in this chapter models only the left heart because of the lack of data and information on the right heart, as well as a perceived general concern for the performance of the left heart, because the left heart performs the greatest amount of work and is usually the first to fail. The following assumptions were made [2]:

1. The left ventricle as a time-varying capacitance
2. The systemic circulation as a four-element Windkessel
3. The pulmonary circulation and left atrium as single capacitances
4. The heart valves as diodes with resistances—the diodes themselves were assumed to be ideal.

The Windkessel effect, meaning "elastic reservoir," is an effect that describes the recoiling effect of large arteries. Deswysen et al. [3] modeled the systemic arterial bed using three elements, as shown in Figure 13.2. The model presented in this chapter makes use of a modified version of the Windkessel effect, which uses an additional elastance component (an inductor), as shown in Figure 13.3.

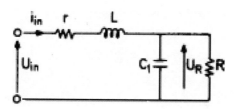

FIGURE 13.3: Four-element arterial bed.

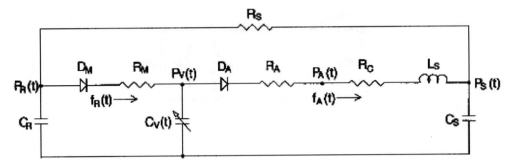

FIGURE 13.4: Analog representation of the left heart cardiovascular model.

Figure 13.4 is an analog representation of the model. Note the variable capacitance C_V, which represents the capacitance of the left ventricle.

The input the model, in Figure 13.4, is the left atrial pressure. From there, the current flows through the mitral diode into the left ventricle. The flow then runs from the left ventricle through the aortic valve modeled as a diode and accompanying resistance. Finally, the flow is seen to pass through the modified Windkessel described above. Output can be considered as the systemic pressure measured at the Windkessel.

The model presents two transfer functions: one for ejection and one for filling of the left heart. The transfer functions are expressed in terms of measurable constants, so that they are verifiable and lend themselves to future research. For ejection, the relevant variables are P_A and f_A. The transfer function G_E can be expressed as:

$$G_E(s) = \frac{f_A}{P_A} = \frac{b_2 s^2 + b_1 s}{a_3 s^3 + a_2 s^2 + a_1 s + 1}$$

where the constants were determined from the following formulae:

$$a_3 = L_S R_S C_S C_R$$
$$a_2 = R_C R_S C_S C_R + L_S(C_S + C_R)$$
$$a_1 = R_S C_R + R_C(C_S + C_R)$$
$$b_2 = R_S C_S C_R$$
$$b_1 = C_S + C_R$$

TABLE 13.1: Averaged values for the constants	
EJECTION	FILLING
$a_1 = 7.146$	$a_1 = 6.781$
$a_2 = 5.715$	$a_2 = 9.924$
$a_3 = 0.521$	
$b_1 = 6.781$	$b_1 = 3.441$
$b_2 = 9.924$	$b_2 = 1.343$

where

$C_V(t)$ = left ventricular compliance, mm Hg/ml

R_C = characteristic resistance, mm Hg s/ml

R_S = systemic resistance, mm Hg s/ml

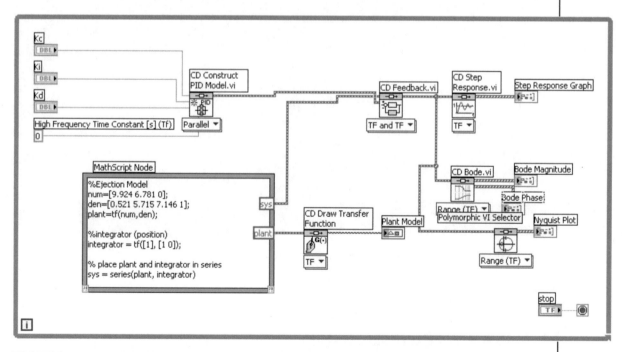

FIGURE 13.5: LabVIEW block diagram for the ejection model transfer function, G_E.

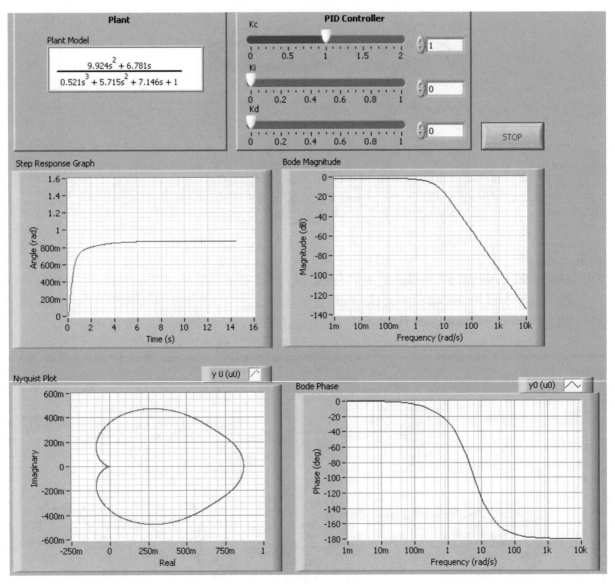

FIGURE 13.6: LabVIEW front panel for the ejection model transfer function, G_E.

L_S = inertance of blood in the aorta, mm Hg s^2/ml

C_S = systemic compliance, mm Hg/ml

C_R = pulmonary circulation, mm Hg/ml

D_R and R_M = mitral valves

D_A and R_A = aortic valves

For filling, the relevant variables are P_V and f_R. The transfer function G_F can be expressed as

$$G_F(s) = \frac{P_V}{f_R} = \frac{-\left(b_2 s^2 + b_1 s + 1\right)}{a_2 s^2 + a_1 s}$$

where the constants were determined from the following formulae:

$$a_2 = R_S C_S C_R$$
$$a_1 = C_S + C_R$$
$$b_2 = R_M R_S C_S C_R$$
$$b_1 = R_S C_S + R_M \left(C_S + C_R\right)$$

Table 13.1 lists the averaged values for the constants used in the above transfer functions.

The model parameters can be identified with P_A and f_A in the ejection phase, and with f_R and P_V in the filling phase. The Laboratory Virtual Instrumentation Engineering Workbench (LabVIEW) block diagram and front panel for the ejection model transfer function, G_E, are shown in Figures 13.5 and 13.6.

The proportional, integral, and derivative (PID) controller was introduced in the feedback loop of the transfer function to introduce some variability with the transfer function. The PID would allow the user in the future to create a set point for the transfer function based on real-world parameters. The default setting for the proportional gain of the PID is set to a value of 1, whereas the integral gain, K_i, and the derivative gain, K_d, are set to 0. An increase in K_i results in the reduction of the operation range of G_E, as in Figure 13.7. An increase in K_d results in the dampening of

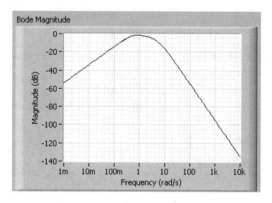

FIGURE 13.7: Increase in K_i results in reduction of operating range.

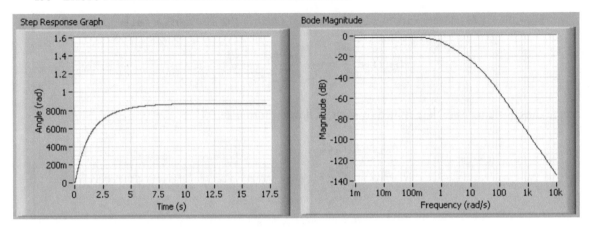

FIGURE 13.8: Dampening of the response with an increase in K_d.

the response shown in Figure 13.8. Note in Figure 13.6 the stability of the system described by the Nyquist plot.

The filling transfer function, G_F, is modeled in LabVIEW as shown in Figure 13.9.

The filling model transfer equation (Figure 13.9) is passed into a CD feedback.vi, in which a PID controller is placed into the feedback loop. The step response, Bode plots, and a Nyquist plot are created. The results can be seen in Figure 13.10.

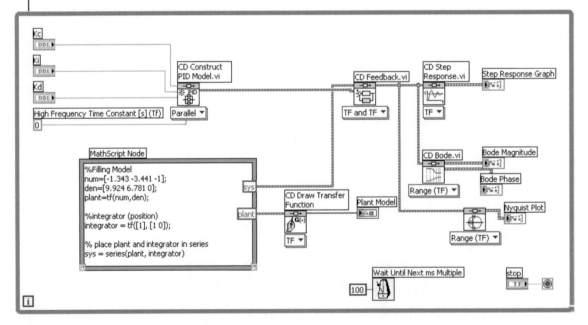

FIGURE 13.9: The filling transfer function, G_F.

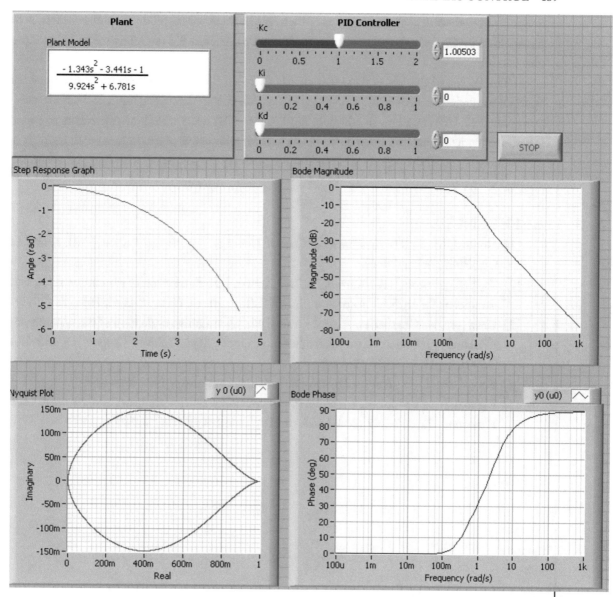

FIGURE 13.10: LabVIEW graphs of the step response, Nyquist, and Bode plots.

Note in the plant equation the negative response of the transfer function exhibited in the numerator. This can be attributed to the negative pressure created when the left heart is filling—as pressure decreases, the filling should increase. Again, a PID controller was introduced in the feedback loop of the system to introduce some controllability in the behavior of the system response. Like in the .vi created for G_E, the proportional gain is set to 1 by default, and the values for the integral and derivative gain are set to zero. Increases in these values produce similar responses to those shown by G_E. In both systems, increases in the k parameters did not affect the stability of the system.

This model of the left heart is particularly useful in the area of left ventricle assist devices (LVAD) research. In creating this model, researchers are able to determine input and output resistances for their LVAD design. In addition, modeling the left heart will allow researchers to better interface a model of their design and its operation thresholds about those in the human heart. It is also possible to model the right heart based on the model of the left heart, with different values for the respective parameters.

REFERENCES

[1] Howard, M. T., *The Application of Control Theory to Physiological Systems*, W.B. Saunders Co., Philadelphia (1966).

[2] Yu, Y. C., Boston, J. R., Simaan, M., and Antaki, J. F., Estimation of systemic vascular bed parameters for artificial heart control, *IEEE Trans. Auton. Control*, **43**(6), 765–778 (1998).

[3] Deswysen, B., Charlier, A. A., and Gevers, M., Quantitative evaluation of the systemic arterial bed by parameter estimation of a simple model, *Med. Biol. Eng. Comput.*, **18**, 153–166 (1980). doi:10.1007/BF02443290

[4] Silverthorn, D. U., *Human Physiology*, Benjamin Cummings, New York (2006).

[5] Elghozi, J. L., and Julien, C., Sympathetic control of short-term heart rate, *Fundam. Clin. Pharmacol.*, **21**, 337–347 (2007). doi:10.1111/j.1472-8206.2007.00502.x

· · · ·

CHAPTER 14

Vestibular Control System

In humans, balance and equilibrium is maintained through a complex control loop involving inputs from the vestibular system, visual system, and proprioceptive sensors. The integration of inputs in the central nervous system (CNS) produces a subjective perception of movement that feeds back to adjust stance and fine motor control. Modeling the maintenance of equilibrium and balance as a control system is virtually impossible using current methods, because of its extreme complexity. This chapter will therefore briefly treat the entire system governing equilibrium and perceived orientation, then focus on the effects of the two vestibular organs: the otoliths and the semicircular canals. To develop a model for these systems, this report will first compile what we know about the physiology of the vestibular system. This information will be used to develop a block diagram control model for both vestibular and otolith control. The time constants and transfer functions for vestibular controls were obtained from the literature, and used to model the vestibular control system in a Laboratory Virtual Instrumentation Engineering Workbench (LabVIEW) Virtual Instruments program.

14.1 PHYSIOLOGY AND ANATOMY
14.1.1 Physiological Basis for Control
Human vestibular control as defined here refers to the maintenance of equilibrium, balance, and orientation using sensory input from the otolithic organs and the semicircular canals that has been integrated and interpreted by the medulla. The vestibular organs are divided both anatomically and by functionality into two distinct systems: the otoliths and the canals [1,2]. Figure 14.1 is a conceptual drawing of the vestibular system.

The otoliths provide sensory input for linear acceleration in the vertical and horizontal planes, which is accomplished through two otolithic organs (the utricle and the saccule), located approximately orthogonal to one another in the inner ear. The utricle is located horizontally in the head and senses accelerations in the horizontal plane, whereas the saccule is located vertically. Both otoliths are composed of a thin membranous sac filled with a viscous gel that suspends small ionic crystals. The membrane is enervated with thousands of tiny hair cells extending into the gelatinous media.

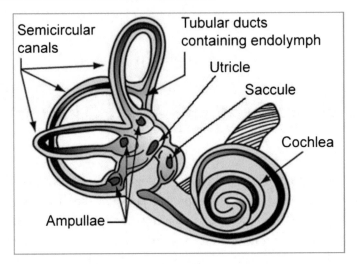

FIGURE 14.1: The vestibular system.

Under a linear acceleration, the stones apply a shear stress to the hair cells, exciting them to increase the number of action potentials per unit time sent through the vestibular nerve to the brain. Under a specific linear acceleration, certain hair cells receive maximum stimulation, whereas hair cells oriented parallel to the direction of acceleration are unstimulated. This directional alignment allows for accurate perception of the direction of acceleration in three dimensions when the separate inputs are integrated in the medulla [1,2]. Figure 14.2 is a conceptual drawing of the otolith organ.

The semicircular canals provide sensory input for angular acceleration in three dimensions (Figure 14.3).

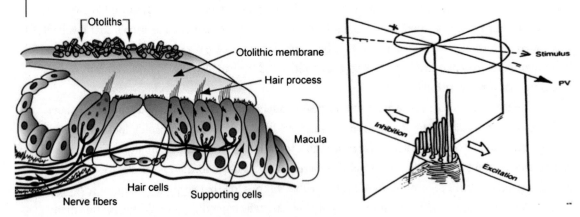

FIGURE 14.2: The otolith organ and directional orientation.

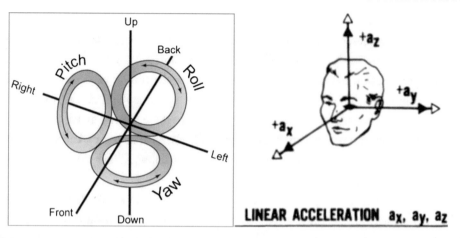

FIGURE 14.3: Orientation and effects on the semicircular canals.

Humans have three semicircular canals on each side of the head oriented orthogonally in roughly the x, y, and z planes (relative to the ground) as shown in Figure 14.3. The mechanism of action for the canals is similar to that of the otolithic organs. Each canal consists of a membranous sac containing a free-flowing viscous fluid. A small raised projection in the center of the canal, called the "cupula," is enervated with thousands of hair cells that connect to the vestibular nerve. Any angular acceleration of the canals will cause inertia to move the hair cells relative to the fluid. Again, an influx of ions in stimulated hair cells will cause an increase in the frequency of action potentials from that particular cell. Integration of the signal strength from all six canals gives a perception of the direction and magnitude of angular acceleration [1,2].

14.1.2 Equilibrium and Balance Control System

It is important to note that equilibrium, balance, and orientation do not rely solely on vestibular input. Visual and proprioceptive input is also integral to maintaining these functions. The visual system provides information about perceived orientation that can be at odds with and even override vestibular input. Just consider the example of sitting in a stationary car while the car next to you begins to move. You have the perception of movement without any vestibular input whatsoever. Proprioceptive input includes the pressure the body feels as a result of the force of gravity. Many times people under water perceive vertical by sensing it with their hands or feet, allowing a mental representation to form based on somatosensory input.

A further complication to the problem of creating a vestibular control model is the fact that the output of the system is a "subjective orientation." This means that two different people with different experiences could perceive a different orientation after exposure to the same stimuli. Humans form a mental representation of what certain actions and movements "feel like," and compare the inputs from the vestibular organs to this mental representation to determine sensation. When the mental representation matches vestibular input, no perception is necessary (we do not even notice). However, when the two do not match, corrective action is required and we become aware of a sensation. As an example, when walking down the street we are not aware of the angular and linear acceleration inputs associated with each step, but when we step on a banana peel corrective action is immediately initiated and we perceive that we are falling. Time constants for the system are dependent on previous experience stored as memory.

This phenomenon is also observed in astronauts subjected to extended space flight. There is an initial adjustment to zero gravity that modifies all of the time constants of vestibular control to better cope with a zero g environment. Upon return to Earth, astronauts must reattain the time constants suitable for terrestrial existence, which they do rather quickly. However, some astronauts have been observed to "revert" to their zero-g configurations when walking on sand or unsteady terrain, resulting in severe vertigo, which is known as latency and is simply a manifestation of the fact that vestibular sensation is based on memory and internal representations as well as external stimuli. Figure 14.4 shows the effect of desired orientation input from the CNS on the semicircular canals. Figure 14.4 is a flow diagram for the equilibrium and balance control.

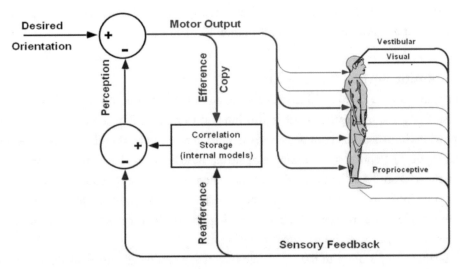

FIGURE 14.4: Information flow diagram for the equilibrium and balance control.

14.2   INTERPRETATION OF BLOCK DIAGRAM
14.2.1 Block Diagram of the Vestibular Control System

The block diagram in Figure 14.5 shows a representation of the vestibular control system. The two inputs of the system are the specific force vector $\bar{f}$ and the angular acceleration vector $\dot{\omega}$. The orientation of the semicircular canals and the otoliths in the head and the instantaneous orientation of the head with respect to inertial space determined the perceptions of these vector inputs on the semicircular canals and the otoliths [3]. Matrix **A** transforms the inertially fixed frame or reference to the frame of reference fixed with respect to the subject's head. The output $\bar{f}_b$ represents the component of specific force along a hypothetical input axis of the otolith. $\dot{\omega}_b$ represents the component of angular acceleration along the input axis of each semicircular canal. These components are multiplied by the sensitivity of the otoliths and canals, and acted upon by the dynamic response of these organs. The CNS is shown to combine the outputs of the otoliths and canals, as well as other sensations, such as visual or tactile, to compose a perceived orientation of man in space. When the vestibular sense is used for closed-loop control such as riding a bicycle or flying an airplane, it leads to manual control compensation by turning a control wheel. This resulting motor control output

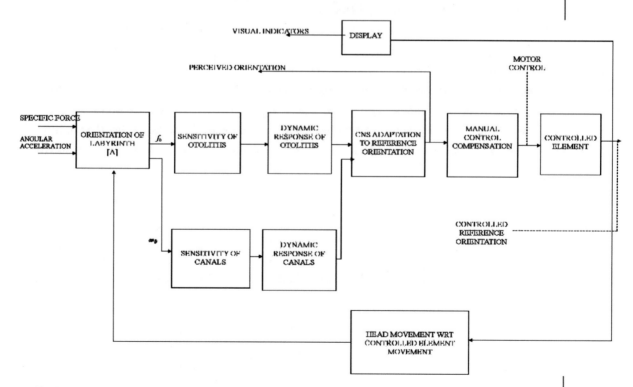

FIGURE 14.5: Block diagram of the vestibular control system.

can be measured as an additional indicator of the controlled reference orientation. Feedback occurs as the orientation of the head in space varies as a direct result of the change in controlled reference orientation. Figure 14.5 is the block diagram of the vestibular control system.

14.2.2 Block Diagram of the Semicircular Canal

The input is assumed to be angular acceleration of the skull with respect to inertial space ω. The matrix transformation ([A]) projects the inertial angular acceleration vector along the input axes of the three semicircular canals. The semicircular canals dynamics shows the cupula output with respect to the input angular acceleration in terms of a highly damped second-order model. The central habituation represents the long time adaptation to successive stimulation of the semicircular canals. The threshold adaptation block may increase the minimum threshold for sensitivity to sensation of rotation and occurrence of nystagmus following a history of angular acceleration. The output of the system is a subjective sensation of angular velocity. Figure 14.6 is the block diagram of the semicircular canal.

14.2.3 Block Diagram of the Otoliths

The orientation of the otolith with respect to the skull and the skull with respect to inertial space yield a set of specific force outputs, $\bar{f}_b$, acting on the otoliths, which is the specific force input of gravity minus the acceleration. The mass M, spring constant K, and damping C of the mechanical arrangement represent the dynamic characteristics of the otolith–macula system. However, the magnitude of these parameters is unknown. The box "position & rate sensitivity" shows the presence of cells whose output is proportional to the change in position of the otolith rather than in its position. The T_L is unknown. The block "directional sensitivity" represents the otoliths' sensitivity to

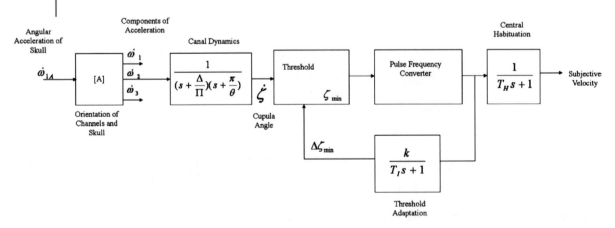

FIGURE 14.6: Block diagram of the semicircular canal.

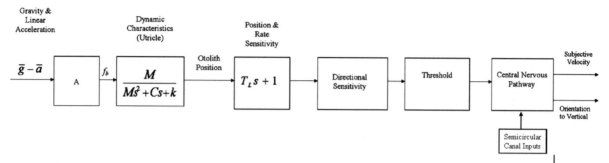

FIGURE 14.7: Block diagram of the otoliths.

changes in the orientation away from the erect position instead of changes in orientation when the subject is tilted at a large angle. Figure 14.7 is the block diagram of the otoliths.

14.3   SIMULATION OF THE CONTROL MODELS IN LABVIEW
14.3.1 Transfer Function of Semicircular Canals
The vestibular system consists of the nonauditory part of the ear, comprising the otoliths and the semicircular canals. The three semicircular canals respond to angular acceleration. The otoliths act in response to linear acceleration as well as gravity. Because both components of the vestibular system are affected by different inputs, it is necessary to model each component individually. In the quest to solve for a transfer function that represented the vestibular system, it became apparent that most, if not all, values are derived through experimental data; to find conclusive values, it was necessary to consider various control models.

The semicircular canal transfer function defines the flow of the endolymph within the semicircular canals as well as the nerve pulse output. To derive the transfer function of the semicircular canals, the movement of the capula was described as a function of head acceleration in time domain (Equation 14.1).

$$K\Theta(t) + B\Theta'(t) + I\Theta''(t) = IH''(t)$$

(14.1)

where

Θ = angular deviation of the endolymph with respect to the skull

K = stiffness or torque moment per unit angular deflection of the cupula

B = moment of friction

I = moment of inertia of the endolymph

H = component of angular acceleration of the skull, with respect to inertial space

The mechanical response can be rewritten as

$$\Theta''(t) = H''(t) - B/I\Theta'(t) - K/\,I\Theta(t) \tag{14.2}$$

It is necessary to convert the equation to frequency domain. Using the Laplace transform changes signals that vary in time to frequency domain by using complex exponentials. Laplace changes the equation to the frequency domain (Equation 14.3).

$$K\Theta(s) + B\Theta'(s) + I\Theta''(s) = IH''(s) \tag{14.3}$$

Using properties of Laplace, Equation (14.4) is obtained.

$$K\Theta(s) + Bs\,\Theta(s) + s^2\,I\Theta(s) = sIH'(s) \tag{14.4}$$

Simplifying Equation (14.4) yields

$$\Theta(s)\left[K + sB + s^2I\right] = sIH'(s) \tag{14.5}$$

Solving for Θ/H the transfer function is obtained with the following equation

$$\frac{\Theta}{H} = \frac{1}{K} x \frac{s}{1 + \dfrac{B}{K}s + \dfrac{1}{K}s^2} \tag{14.6}$$

Then, by combining time constants, the following transfer function is obtained using Equation (14.7)

$$\frac{\Theta}{H} = \frac{T_1 T_2 s}{(T_1 s + 1)(T_2 s + 1)} \tag{14.7}$$

where T_1 is the short time constant and T_2 is the cupular time constant, and

$$T_1 + T_2 = \frac{B}{K}.$$

Values for time constants were estimated to as: $T_1 = 0.006$ s and $T_2 = 13$ s. As soon as the transfer function equation was obtained, a model of the semicircular canals was created in Lab-VIEW (Figure 14.8). A MathScript node was used to define the numerator and denominator of the canal transfer function. To the MathScript node, controls for T_1 and T_2 were implemented as

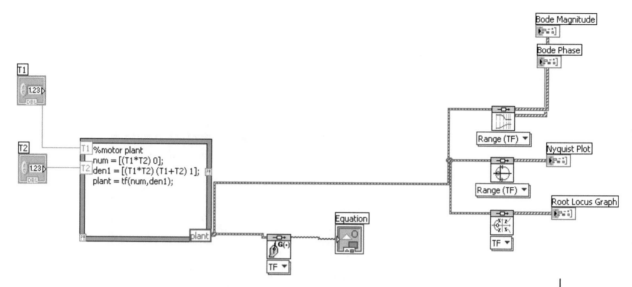

FIGURE 14.8: LabVIEW block diagram of the semicircular canal transfer function.

inputs. The output plant is wired to a CD Draw Transfer Function Equation.vi, CD Bode.vi, CD Nyquist.vi, and a CD Root Locus.vi. The CD Draw Transfer Equation.vi has an equation indicator, which draws a picture of the model equation. The CD Bode.vi produces the Bode magnitude and Bode phase plots of the system. The CD Nyquist.vi produces a Nyquist plot, which plots the imaginary part of the frequency part of the frequency response to its real part, and the CD Root Locus.vi plots the root locus.

The frequency response of the semicircular canals describes the relationship between the head movement and the magnitude of the endolymph movement. The phase of the response measures the timing with respect to the stimulus waveform. A response that rises and falls in time with the stimulus is in phase with the stimulus; one that leads or lags behind the response is not in phase with the stimulus. It is important to note that the transfer function derived is for the endolymph displacement as a function of head velocity. From the Nyquist plot, we see that the system is stable, because the Nyquist diagram does not encircle the −1 point.

Figure 14.9 presents from top to bottom the LabVIEW front panel window, which shows the resulting log magnitude and phase graphs (Bode plots), the Nyquist graph, and the root locus graph for the semicircular canal transfer function.

The transfer function obtained for the otolith was derived using Trincker (1962) [4] equations describing the mathematical response of the otolith to applied acceleration. The Trincker equation is as follows:

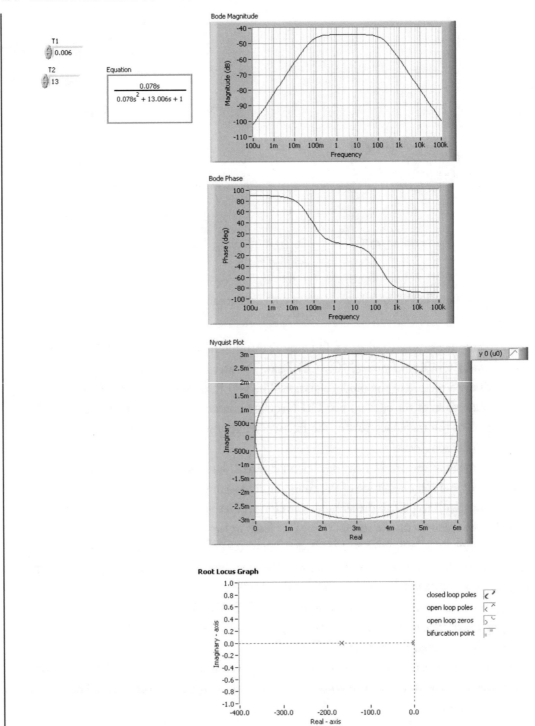

FIGURE 14.9: LabVIEW front panel of semicircular canal transfer function results.

$$\left(1 + \frac{1}{\rho}\right) m\ddot{x}_o + r\dot{x}_o + kx_o = \left(1 - \frac{1}{\rho}\right) m\ddot{x}_n$$

$$(14.8)$$

where

m = mass of the otolith

r = viscous force per unit linear velocity

k = elastic restoring force per unit linear displacement

ρ = density of the otolith

x_o = unit linear displacement of otolith with respect to head

x_n = unit linear displacement of the head

Making assumptions of physiological transduction, Equation (14.8) can be expressed as a transfer function relating acceleration to impulse frequency in the first-order afferent neuron. It can be said that impulse frequency was a function of both otolith displacement, x_o, and otolith velocity, x_o', as shown by Equation (14.9).

$$R_\Gamma = f_1(x_o) + f_2(\dot{x}_o)$$

$$(14.9)$$

Combining both equations and expressing the result in Laplace notation derives the transfer function shown in Equation (14.10).

$$H(s) = \frac{k_1 + k_2 s}{(s + w_1)(s + w_2)}$$

$$(14.10)$$

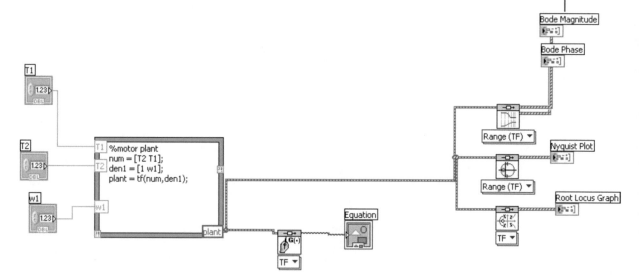

FIGURE 14.10: LabVIEW block diagram of the transfer function of otoliths.

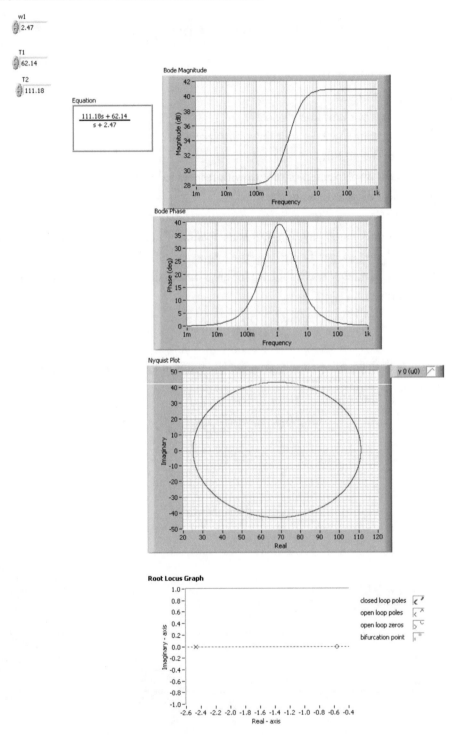

FIGURE 14.11: LabVIEW front panel window for the transfer function of otoliths.

where

k_1 = gain with respect to otolith displacement expressed in impulses

k_2 = gain with respect to otolith velocity expressed in impulses

s = Laplace operator

w_1 = lower corner frequency

w_2 = upper corner frequency

The following values—k_1 = 62.14 imp cm^{-1}, k_2 = 111.18 imp cm^{-1}, w_1 = 2.47 rad s^{-1}— were used in the transfer function to simulate the model in LabVIEW; and w_2 can be ignored at high frequencies. The transfer function relates the total response, R_T, to acceleration of the head, dx^2_n/dt^2.

Figure 14.10 shows the block diagram of the otolith transfer function used to program Equation (14.11) in LabVIEW controls. The LabVIEW sub-VIs used in the program include A Math-Script node, the CD Draw Transfer Function Equation.vi, the CD Bode.vi, the CD Nyquist.vi, and a CD Root Locus.vi.

$$G(s) = \frac{111.18s + 62.14}{s + 2.47} \qquad (14.11)$$

Figure 14.11 presents from top to bottom the LabVIEW front panel window, which shows the resulting log magnitude and phase graphs (Bode plots), the Nyquist graph, and the root locus graph for the otolith organ transfer function.

REFERENCES

[1] Guyton, A. C., *Textbook of Medical Physiology* 4th Ed., W. B. Saunders Co. Philadelphia, PA (1971).

[2] Boron, W. F., and Boulpaep, E. L., *Medical Physiology*, Elsevier Saunders, Philadelphia, PA (2005).

[3] Highstein, S. M., Fay, R. R., and Popper, A. N., *The Vestibular System*, Springer, Berlin (2004).

[4] Trincker, D., The transformation of mechanical stimulus into nervous excitation by the labyrinthine receptors. Symp. Soc. Exp. Biol. 16:289 (1962).

CHAPTER 15

Vestibulo-Ocular Control System

The function of the vestibulo-ocular reflex (VOR) is to stabilize an image on the surface of the retina during head movement. The gain of the VOR is the eye velocity/head velocity ratio, which is typically close to 1 (approximately 0.95) when the eyes are focused on a distant target. However, to stabilize images accurately, the VOR gain must vary with context. The balancing system comprises two complex processes that can be grouped into gaze stabilization and postural stabilization. The vital components of gaze stabilization are VOR, smooth pursuit (SP) system, and saccadic eye movement. The kinematic model of the VOR, which relies on sensory information available from the semicircular canals (head rotation), the otoliths (head translation), and neural correlations of eye position, is described. Using the Laboratory Virtual Instrumentation Engineering Workbench (LabVIEW) Control and Simulation Module, a computational model was developed and programmed to produce the amplitude and time course of the VOR modulation. The steady-state gain of the system is modified by changing the ratio of the two time constants along the feed-forward and the feedback projections to the Purkinje cell unit in our model VOR network. Our analysis thus provides a thorough characterization of the system and could thus be useful for guiding further physiological tests of the model. In summary, this chapter describes the VOR in some detail, from the mathematical and physiological perspectives, and presents a computational model of how this topographic organization can be deduced from the information presented to the structure.

15.1 STIMULUS

The stimulus for the VOR is head acceleration. When the head moves, the VOR responds with an eye movement that is equal and opposite in direction. With a rotational movement, the head moves relative to the body. Translational movements occur when the entire body is moved in tandem. Rotational VOR responds to angular motion of the head and results from the stimulation of the semicircular canals, whereas translational VOR responds to linear motion of the head and results from stimulation of the otolithic organs. Some head movements may involve a combination of both translational VOR and rotational VOR.

15.2   RESPONSE

VOR attains stabilization of the object in the visual field by controlling the eye muscles to compensate for the head acceleration [1]. The balance system includes a complex array of control processes that can be grouped into two distinct, but interdependent, systems. One is the gaze stabilization system, which maintains gaze direction of the eyes and visual acuity during activities involving active head and body movements. The second one, the postural stabilization system, keeps the body in balance while an individual stands and actively moves about in daily life. Both the gaze stabilization and postural stabilization system are distinct because they rely on information from different senses, motor reactions of different parts of the body, and are initiated by different brain pathways. However, both systems are interdependent because gaze stability cannot be achieved unless the body on which the head and eyes ride is also stable.

15.3   NORMAL PERFORMANCE

15.3.1 Saccadic Eye Movements

Stability of gaze can be maintained by relying on interactions among combinations of gaze direction information, groups of eye muscles, and the brain's ability to integrate the sensory and motor functions. It is well known that when an object is noticed by the eyes, the eyes are rapidly turned toward it. The same ocular motor response can be provoked by a sudden noise or a stimulus applied to the body's surface. This rapid eye movement is called a saccade. Saccades occur at a rate of about 3 saccades/s in the alert state, and the majority are very small, e.g., when reading or looking at a face. The saccadic eye movement system generates rapid eye movements when the VOR and SP systems fail to maintain gaze on a desired visual target. Saccadic movements require the presence of a visual target but are unaffected by vestibular information. The latency period of the saccadic system is around 200 ms.

15.3.2 Smooth Pursuit System

When the object is moving, the saccadic system is unable to hold on to the image for a long time as the image slides off the fovea; however, the SP system overcomes this deficiency and smoothly tracks a target. The fovea is that part of the retina where the visual activity is the best. If the target moves too fast for SP, the saccadic system is activated to recapture it. The latency period for a pursuit movement is usually 100–130 ms. When target movements are repeated and become predictable to the subject, then the accuracy and timing of SPs are much improved. The SP system requires the presence of a visual target but is not affected by the presence or absence of vestibular information. Our model was based on this system (discussed in subsequent sections).

15.3.3 Vestibulo-Ocular Reflex and Vestibulo-Collic (Closed-Loop VCR) Reflexes

When SP stabilizes the image of a moving object or a target on the fovea, VOR is required to stabilize the fixation of a stationary target during head rotation. The latency period of VOR is about 10 ms. When a person is walking, vibrations from the heel are transmitted to the head. That is when the combination of VOR and VCR stabilizes the visual activity. The gain of the system is defined as the ratio of the eye velocity to the head velocity, which is close to 1 under normal conditions. In these terms, the gain of the VOR in normal mammals is very close to 1 even in darkness at head speeds of up to 300°/s because of its dependence on vestibular rather than visual stimuli.

To explain VOR action better, the following experiment may further illustrate the independency of VOR from visual stimuli. First, keep your head facing in one direction, and move your hand fairly quickly backward and forward in front of you, trying to track only with our eyes. The image of your hand will be blurry. Next, keep your hand still and move your head from side to side. Even when the speeds are about the same, the image of your hand is much clearer under this condition.

The VOR system is a fast-acting system in comparison to SP that relies on inputs from the vestibular system to drive eye movements in equal and opposite direction to the direction of the head. The gain of the VOR movement is influenced by the subject's state of arousal. The VOR is effective for stabilizing gaze during rapid movements" and is ineffective for slow movements.

15.4   PHYSIOLOGICAL PATHWAYS

When taking into account the response of the vestibular system, there are several distinct pathways that have an effect on the vestibular ocular reflex. Each of these pathways is considered intricate, but all the pathways mentioned throughout this paper have a direct effect on the physical response of the eye. It is important to note that although specific pathways are mentioned in this chapter, a single stimulus may also have an effect on pathways that may not be mentioned or accounted for in this chapter, but which are still relevant.

There are four distinct vestibular nuclei located in the brain stem to consider when observing the vestibular response of the eye to an initial stimulus. All cellular types within the brain stem (superior, lateral, and posterial vestibular nuclei) receive their primary input from other portions of the central nervous system. This chapter will be primarily concerned with developing a LabVIEW program for analyzing the vestibular response to a change in head position or rapid head movement.

The vestibular semicircular canal (Figure 15.1) contains the ampulla, which in turn contains a crista or transversely oriented ridge of excess tissue. In addition, the ampulla contains ciliated hair cells that are used to sense movement of the gelatinous fluid surrounding the hair cells. All hair cells are oriented in random directions to detect the movement of the endolymphic fluid in any direction. This essentially allows a group of hairs to produce a stimulus whereas other hairs oriented in a different direction produce fewer stimuli. The release of neurotransmitter by hair cell vesicles causes an increased firing rate of the action potential within the vestibular ganglion, producing a signal that continues on to the vestibular nuclei of the brain stem. Essentially, the rotation of the head to the right stimulates the hair cells in the crista of the right horizontal canal and inhibits the left horizontal canal. This stimulus results in the stimulation of the right vestibular nerve and consequently produces a reflex of the ocular system in the opposite direction of horizontal angular acceleration.

In addition to the semicircular canals located on each side of the brain, the otolith associated with each of these canals also creates a similar stimulus as the semicircular canals. Although the otolith organs (saccule and utricle; Figure 15.1) produce a similar stimulation to head movement as the semicircular canals, their response is produced by linear acceleration instead of angular acceleration. Both of these organs also contain patches of aligned hair cells called the macula that are stimulated by the movement of hexagonal prisms of calcium carbonate called otoconia. Because the otoconia of the utricle and saccule are considerably denser than the endolymph, the otolith membrane will be displaced by gravity or other linear accelerations instead of angular accelerations. This displacement causes an increased or decreased stimulus, in turn releasing neurotransmitter in a similar respect as the semicircular canals.

Stimulation of the vestibular nuclei via the release of neurotransmitter is used in the maintenance of balance and the stabilization of visual images on the retina during head movements. This, in turn, causes a signal output to the paramedian pontine reticular formation (PPRF), which lies within the medial portion of the pontine tegmentum. Neurons in their respective PPRF stimulate

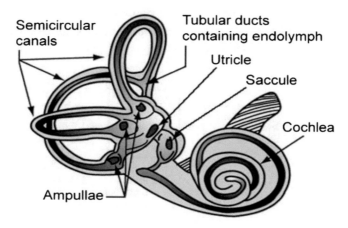

FIGURE 15.1: The vestibular system. The semicircular canal consists of three circular tubes lying in mutually perpendicular planes containing endolymph that are capable of detecting head movement in any direction via a combination of the three canals.

the abducens nucleus involved in the generation of horizontal eye movements. The abducens contains two types of neurons: the large motor neurons that pass ventrally through the pons to exit on the surface of the brain stem, and smaller neurons that do not leave the brain stem but cross the medial longitudinal fasciculus (MLF). The smaller neurons stimulate oculomotor nuclei of the opposite eye, resulting in the stimulation of the medial rectus of the opposite eye. In addition, the abducen directly stimulates the lateral rectus of its respected eye. Consequently, the two stimuli, in conjunction with one another, cause the simultaneous horizontal eye movement of the eyes in a single direction. Essentially, a quick movement of the head to the right causes stimulation of the right vestibular nuclei resulting in the stimulation of the left PPRF, the left abducen, the left lateral rectus, the right oculomotor nucleus, and the right medial rectus [1]. The physiological pathway block diagram is shown in Figure 15.2.

As indicated, stimulation of the left vestibular nuclei also has an effect on a smaller descending pathway called the medial vestibulospinal tract or MLF. Cells within this pathway possess axons that descend in a position near the dorsal surface of the pons and medulla. Connections are also made with the cervical and upper thoracic motor neurons, which both play a role in the normal positioning of the neck. The movement of your head or neck in a direction causes stimulation of the medial vestibular nuclei, informing higher-order brain function of the movement leading to compensation of the erect head.

The voluntary turning of both eyes in the horizontal direction to view a desired object is not accomplished by any of the pathways mentioned above, but rather is completed by the stimulation of the PPRF by the front eye fields (FEFs). This movement is often referred to as left or right horizontal saccade or jerking. In similar respects to the involuntary movement of the eye via the otolith and semicircular canals, the FEFs produce stimuli that result in the simultaneous movement of both eyes via the lateral and medial rectus. As shown in Figure 15.3, the stimulus originates in the frontal

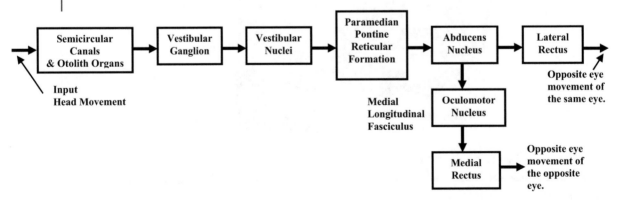

FIGURE 15.2: Semicircular canal block diagram.

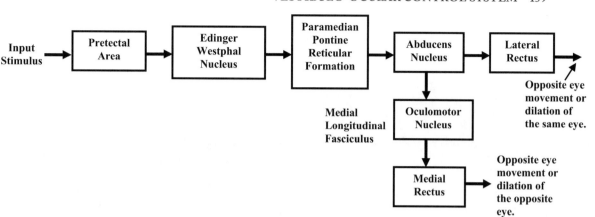

FIGURE 15.3: Block diagram of the front eye field pathway.

eye fields of the cerebral cortex and proceeds to the fretectal area and, ultimately, Edinger–Westphal nuclei. This complex cellular unit receives retinal input and is the essential component in the compensation of the pupil dilation resulting from an increase or decrease in light exposure. In addition, the Edinger–Westphal nucleus is used in focusing and accommodation of the eye, and also has a similar resulting pathway following the stimulation of the PPRF.

In addition to the previously mentioned pathways, the center for lateral gaze also has a direct affect on the vestibular nuclei of the brain stem. When taking into account the movement of the eye, there are two movements to consider: horizontal and vertical gaze. The exact control and movement of the eye requires input from multiple centers in the brain that control the oculomotor, trochlear, and abducen nuclei. Essentially, the combination of these three nuclei controls the six muscles used to control eye movement (Figure 15.4). When considering eye movement in the vertical plane, there has to be exact movement and stimulation of the superior rectus, inferior rectus, inferior oblique, and superior oblique [4]. The center for vertical gaze coordinates the movement of each of these muscles by stimulation of oculomotor nuclei and trochlear nuclei. In addition to torsional movement, there is also control by the stimulation of each of the muscles together, via the elevation and depression of the eye.

There are several physiological pathways that exert a direct and an indirect effect on the VOR as discussed in detail above. In addition to each of these primary pathways, the vestibular nuclei are also affected by the additional feedback loops of higher-order brain functions such as the dorsolateral pontine nucleus, floccules lobe, and the medial superior temporal, which receives information for optic flow [3]. As depicted in the overall physiological block diagram, shown in Figure 15.5, some of these pathways overlap and integrate with each other to produce a distinct response.

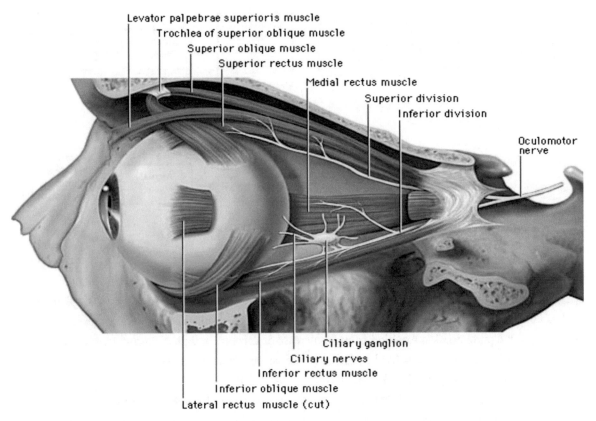

Levator palpebrae superioris muscle
Trochlea of superior oblique muscle
Superior oblique muscle
Superior rectus muscle
Medial rectus muscle
Superior division
Inferior division
Oculomotor nerve

Ciliary ganglion
Ciliary nerves
Inferior rectus muscle
Inferior oblique muscle
Lateral rectus muscle (cut)

FIGURE 15.4: Representation of the eye muscles [4].

The VOR is defined as the compensatory eye movement for any head rotation or movement. This keeps the visual image fixed on the retina, allowing us to design a representation of the ocular response to head movement. Essentially, by defining certain parameters, Coenen and Sejnowski [2] were able to mathematically formulate a response for a given head movement. Using the diagram shown below, they were able to define the resulting Equation (15.1) for complementary resulting velocity.

$$\vec{\omega} = -\vec{\Omega}_c + \frac{\hat{g}}{|g|} \, x \left[\vec{o}_j \vec{\varepsilon}_j \, x \vec{\Omega}_c - \vec{T}_{oj} \right] \qquad (15.1)$$

where

$\vec{\Omega}_c$ = head rotation velocity sensed by the semicircular canals

$\vec{T}_{oj}$ = head translation velocity sensed by the otolith

$\vec{\varepsilon}_j$ = constant vector specifying the location of an eye in the head

$\vec{o}_j$ = position of either the left or right otolith

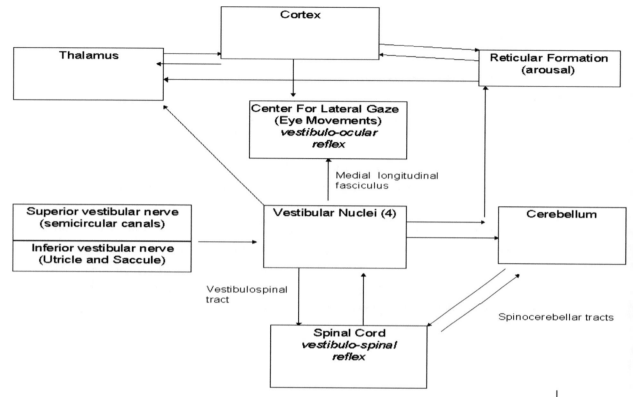

FIGURE 15.5: Overall physiological block diagram for VOR.

$\hat{g}$ and $|g|$ = unit vector and amplitude of the gaze vector

x = cross-product between two vectors

$\vec{\omega}$ and $\vec{\Omega}_c$ = rotation vectors describing the instantaneous angular velocity of the eye and head

For this particular model, a rotation vector lies along the instantaneous axis of rotation. The magnitude indicates the speed of rotation around the axis, and its direction is given by the right-hand screw rule. The combination of a rotation velocity measured by the semicircular canals and a translation velocity is sensed by combining these two velocities. Essentially, the rotation vectors are equal, and the translation velocity vectors are measured by equation described by the otolith given by

$$\vec{T}_{oj} = o\vec{a}_j x\vec{\Omega} + \vec{T} \qquad (15.2)$$

where

$o\vec{a}_j \equiv (\vec{a} - \vec{o}_j)$

$\vec{a}_j$ = position vector of the axis of rotation

Considering the special case where the gaze is horizontal and the rotation vector is vertical, ω can be simplified by writing its equation with the resulting dot product. Because $\hat{g}$ and $\vec{\Omega}_c$ are considered to be perpendicular, the resulting dot product of the two is zero, which results in the first term of the following expression (Equation 15.3) in brackets to be zero [5].

$$\vec{\omega} = -\vec{\Omega}_c + \frac{1}{|g|}\left[o\vec{\varepsilon}\left(\hat{g}\cdot\vec{\Omega}_c\right) - \vec{\Omega}_c(\hat{g}\cdot o\vec{\varepsilon}) - \bar{g}x\vec{T}_o \right] \qquad (15.3)$$

Recall that the semicircular canals report acceleration and velocity of head rotation by its components along the three perpendicular canals on each side of the head. In addition, the saccule and utricle on each side report the linear acceleration, which means that we need to determine the physiological source of both gaze vector (g) values. The eye position is assumed to be provided by the output neural integrator, providing eye position information that is necessary for the activation of neurons to sufficiently focus the eye in a fixed position. Taking this information into account, assume the eye position to be the coordinates of the unit vector along vectors l_1 and l_2 as shown in Figure 15.6. In addition, the gaze vector (g) value also depends on the eye velocity as given by

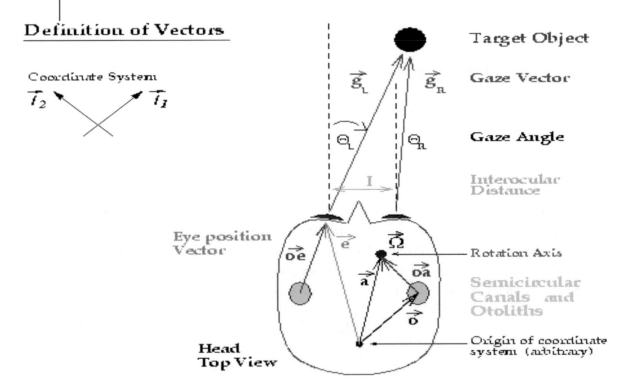

FIGURE 15.6: Kinematic model of the eye position with unit vectors l_1 and l_2 coordinates.

$$\frac{d\vec{g}}{dt} = \hat{g}x\vec{\omega}$$

$$\vec{\omega}(t) = \omega(t)\hat{z}$$

(15.4)

15.5 SPECIAL CASE

There is a special case where the eye position coordinates are defined by the following equations:

$$\hat{g}_1(t) = \hat{g}_1(0) + \int_0^t \hat{g}_2(\tau)\omega(\tau)d\tau$$

$$\hat{g}_2(t) = \hat{g}_2(0) + \int_0^t \hat{g}_1(\tau)\omega(\tau)d\tau$$

(15.5)

These equations are essentially a set of two negatively coupled integrators, which leads us to believe that the neural integrator does not integrate the velocity of the eye directly, but rather integrates it as a product of eye position and eye velocity together. In addition, the distance from the eye to the target can be defined using the gaze angles in the horizontal plane of the head as given by Equations (15.6) and (15.7).

$$\text{Right eye:} \quad \frac{1}{|g_R|} = \frac{\sin(\theta_R - \theta_L)}{I\cos(\theta_L)} = \frac{1}{I}\sec(\theta_L)\sin(\theta_R - \theta_L)$$

(15.6)

$$\text{Left eye:} \quad \frac{1}{|g_L|} = \frac{\sin(\theta_R - \theta_L)}{I\cos(\theta_R)} = \frac{1}{I}\sec(\theta_R)\sin(\theta_R - \theta_L)$$

(15.7)

where

$\theta_R - \theta_L$ = vergence angle

I = interoccular distance (the measured angles are from a directly forward gaze and negative values are used when the eyes are turned toward the right)

Moving from the kinematic model, consider the physiological models for computational analysis by Young and Stark [8].

15.6 COMPUTATIONAL MODEL

15.6.1 Traditional Model: Young and Stark Model

One of the earliest models that dealt with computational modeling of the ocular vestibular system is the Young and Stark model. In 1963, Young and Stark [8] developed a model of the saccadic eye movement system; however, at that time not much was known about the physiological pathway

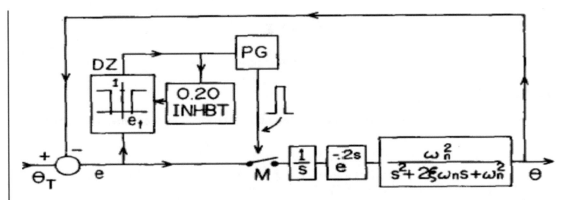

FIGURE 15.7: The Young and Stark model of the ocular vestibular system.

and as such they ignored physiological integrators and came up with the pathway shown in Figure 15.7 [8].

In this pathway, θ_T denotes eye position, S is stimulus rate, s is the Laplace transform variable, θ is the target position, ω_n is the neper frequency (240 rad/s), ξ is damping ratio (0.7). Young and Stark [8] incorporated a negative feedback control as INHBT, giving a delay of 0.20 s.

The Young and Stark model was not used in the development of a LabVIEW analytical model because the saccadic eye moment involves a nonlinear transfer function, and the Young and Stark model solves an underdamped response (whereas it is known that the human body's response is overdamped). So that the target position of the eye can be reached quickly and with as little disturbance as possible, human response is always overdamped. Also, the Young and Stark model of the ocular vestibular system solves for a stationary target, but does not predict the response when the target is moved in ramp or step ramp motion. In addition, the system did not incorporate physiological integrators, because at that time, the physiological pathway for the ocular system was not solved. Young and Stark [8] used linear regression as a means to further derive the transfer function. Because it is known that not many things in the human body are linear, it was decided not to use the Young and Stark model for the LabVIEW computational analysis.

15.6.2 LabVIEW Computational Analysis with the Lisberger–Sejnowski VOR Model

A physiological pathway VOR model developed in 1992 by Lisberger and Sejnowski was conducted using monkeys [6,7]. This physiological pathway was used to create the block diagram shown in Figure 15.8.

This pathway solves for a positive and a negative feedback loop, namely, the subcortical optokinetic pathway and the cortical optokinetic pathway. Figure 15.8 deals with several nonlinear

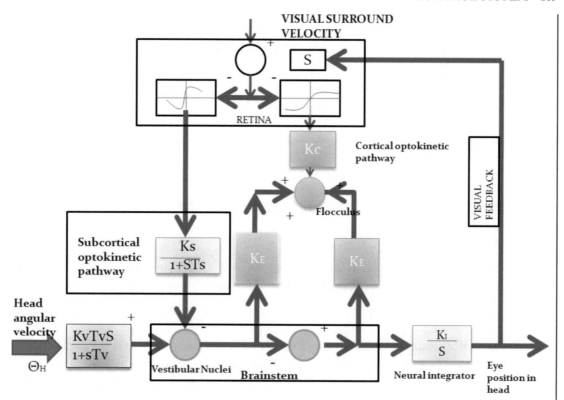

FIGURE 15.8: Block diagram of the physiological pathway in the VOR model.

responses. The retina gives a nonlinear response as do all the physiological units. For simplicity, it was decided to consider the physiological responses as a black box each representing a unit in the pathway given in Figure 15.8. To solve for the transfer function ignoring the nonlinearities and solving only for the linear portion of the response, a simplified version of Figure 15.8 was used, as shown in Figure 15.9.

This transfer function is then solved by using the linear form

$$o(t) = i(t) * e^{-\frac{t}{\tau}} \qquad (15.8)$$

where
* = denotes convolution
$i(t)$ = total input to the unit
$o(t)$ = total output from the unit
G = time constant of the unit

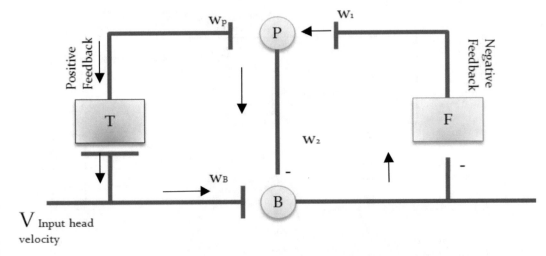

FIGURE 15.9: Simplified version of the block diagram for the VOR physiological pathway. B, brain stem; T and F, relay stations; P, Purkinje fibers in the cerebellum; W, weight of synaptic interactions.

The VOR system was analyzed using frequency domain manipulation via the Laplace transform, which converted the convolutions into multiplications. To represent the Laplace transform form of the physiological units, the terms $T(s)$, $P(s)$, $B(s)$, and $W(s)$ were calculated via Equations (15.9), (15.10), (15.11), and (15.12), respectively.

$$T(s) = V(s)\frac{1}{s\tau_T + 1} \tag{15.9}$$

$$P(s) = [W_P T(s) + W_1 F(s)]\left(\frac{1}{s\tau_P + 1}\right) \tag{15.10}$$

$$B(s) = [W_B V(s) + W_2 P(s)]\left(\frac{1}{s\tau_B + 1}\right) \tag{15.11}$$

$$F(s) = - B(s)\frac{1}{s\tau_F + 1} \tag{15.12}$$

where
    $V(s)$ = Laplace transform of input head velocity
    G = time constant relating to the particular type of physiological unit
    s = Laplace transform
    $T(s)$ = Laplace transform of positive feedback loop
    $P(s)$ = Laplace transform of purkinje fiber response

$B(s)$ = Laplace transform of brain stem

$F(s)$ = Laplace transform of negative feedback loop

The next step is to solve for each one of the physiological units. Equation (15.13) is used to solve for the brain stem unit, $B(s)$:

$$B(s) = H(s)V(s) \qquad (15.13)$$

where $H(s)$ is solved via Equation (15.14), because $V(s)$ is simply the Laplace of the input function.

$$H(s) = \frac{(s\tau_F + 1)\left\{W_B\left[s^2\tau_T\tau_P + s(\tau_T + \tau_P)\right] + W_B - W_2 W_P\right\}}{(s\tau_F + 1)\left[s^3\tau_B\tau_F\tau_P + s^2(\tau_B\tau_F + \tau_F\tau_P + \tau_P\tau_B) + s(\tau_B + \tau_F + \tau_P) + 1 - W_1 W_2\right]} \qquad (15.14)$$

Equation (15.14) solves for the brain stem response to rapid eye movement (saccadic eye movement), SP (slow eye movement), and VOR; however, the saccadic eye movement is mostly nonlinear as is the VOR. The SP has a substantially important linear portion; therefore, it was decided to solve the equation for SP.

Smooth pursuit. To solve for linear portion of SP, the following assumptions must be considered:

1. $W_P = 0$.
2. W_B represents the total synaptic interaction.
3. W_B' represents only the linear portion of the synaptic interaction.

By substituting these assumptions into Equations (15.13) and (15.14), Equations (15.15) and (15.16) are obtained.

$$B(s) = H_{SP}(s)U(s) \qquad (15.15)$$

$$H_{SP} = \frac{W_B'(s\tau_F + 1)(s\tau_P + 1)}{(s\tau_B + 1)(s\tau_F + 1)(s\tau_P + 1) + W_B'(s\tau_F + 1)(s\tau_P + 1) - W_1 W_2} \qquad (15.16)$$

For the LabVIEW computational modeling, the following numerical assumptions were made to configure the program to graph the SP model.

1. $G_F = G_T = G_B = G_P = 100$ ms
2. $W_B' = 0.9$
3. $W_1 W_2 = 1$

The Lisberger–Sejnowski VOR model [6,7] was formulated using monkeys; hence, the block diagram and the values used in this model represent the physiological pathway obtained from monkeys. The numerical values used by the authors in this chapter for the LabVIEW computational model represent values used for the human VOR; thus, use of these values results in the transfer function for humans as described by Equation (15.17).

$$H(s) = \frac{9(s + 10)^2}{(s^3 + 39s^2 + 480s + 900)}$$ *(15.17)*

15.7   RESULTS OF THE LABVIEW ANALYSIS

The log magnitude and the phase graphs (called "Bode plots") have the characteristic of a low pass filter, which is how SP may be modeled. Because SP deals with slow eye movements that only works at low frequencies, the curves resemble the response one might expect from a low pass filter. The LabVIEW front panel shows the resulting log magnitude and phase graphs (Bode plots) in Figure 15.10.

According to the Nyquist criteria, "A closed loop that does not encircle the negative one (−1) on the real axis represents a stable system"; therefore, the SP (slow eye movement) for humans as described by Equation (15.17) is a stable system. In Figure 15.11, take note of the equation shown in the upper right-hand side of the LabVIEW front panel window.

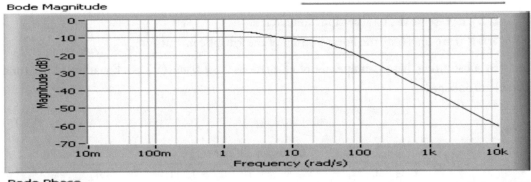

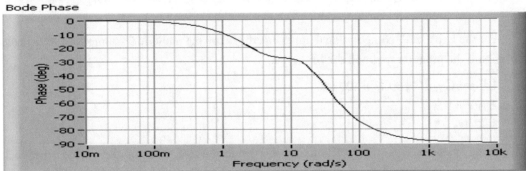

FIGURE 15.10: The log magnitude and phase graphs for smooth pursuit.

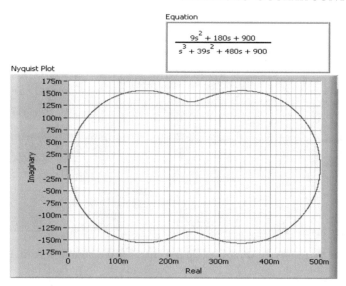

FIGURE 15.11: Nyquist plot for smooth pursuit.

Figure 15.12 shows the LabVIEW results of the root locus plot for SP. A system is said to be stable if all its poles are in the left-hand side of the s plane. Note in Figure 15.12 that at the operating point, the system described by Equation (15.17) is considered to be stable, because all of the poles lie in the left-hand side of the s plane in the root locus plot.

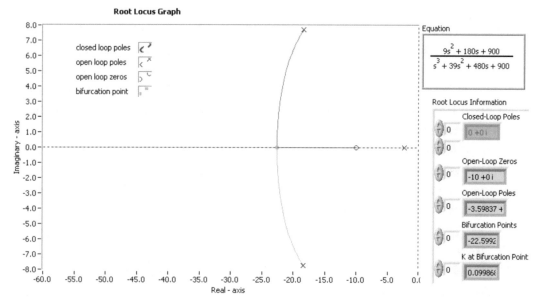

FIGURE 15.12: LabVIEW results showing the root locus graph for the human smooth pursuit system.

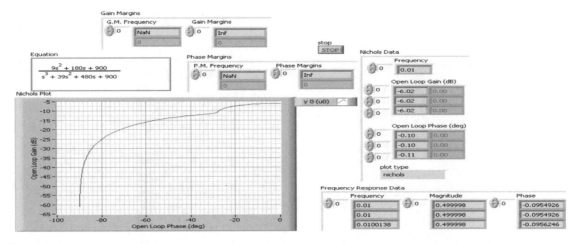

FIGURE 15.13: LabVIEW results showing the Nichols chart for the human smooth pursuit system.

Figure 15.13 shows the LabVIEW results of the Nichols chart for SP. The Nichols chart combines the information from the log magnitude and phase graphs, and plots the gain (log magnitude) information along the *y* axis and the phase information along the *x* axis of a single graph. As in the previous graphs, the Nichols chart also indicates that the human SP system described by Equation (15.17) is stable.

15.8   SUMMARY

During motion in a stationary visual surround, visual stabilization can be obtained if head rotations are compensated for by eye rotations in the opposite direction. The vestibular system perceives head rotations and the visual system detects the slip of the image on the retina. The interaction between these two sensory systems is organized by a reflex pathway called VOR. The LabVIEW results of the SP model indicated a stable, overdamped response to low-frequency movements of the eye, which is consistent with available physiological data. It should be noted that the LabVIEW computational model only solved for the linear part of the equations, and that the model assumed human values for a physiological system that was solved for monkeys. Hence, one can expect that there will be discrepancies in the transfer functions as well; however, considering all of the assumptions, the computational model worked as well as expected.

REFERENCES

[1]   Burdess, C., The Vestibulo-Ocular Reflex: Computation in the Cerebellar Flocculus, http://bluezoo.org/vor/vor.pdf, June 1996.

[2]   Coenen, O. J., and Sejnowski, T. J., A Dynamical Model of Context Dependencies for the Vestibulo-Ocular Reflex, http://www.cnl.salk.edu/~olivier/nips95submhtml.html, February 1996.

[3]   Raymond, J. L., and Lisberger, S. G., Neural learning rules for the vestibulo-ocular reflex, *J. Neurosci.*, **18**(21), 9112–9129 (1998).

[4]   Cranial Nerve III—Oculomotor Nerve, Yale University School of Medicine, http://www .med.yale.edu/caim/cnerves/cn3/cn3_7.html, March 1998.

[5]   Buizza, A., and Schmid, R., Visual–vestibular interaction in the control of eye movement: mathematical modeling and computer simulation, *Biol. Cybern.*, **43**(3), 209–223 (1982). doi:10.1007/BF00319980

[6]   Qian, N., Generalization and analysis of the Lisberger–Sejnowski VOR Model, *Neural Comput.*, **7**(4), 735–752 (1995).

[7]   Lisberger, S. G., The neural basis for motor learning in the vestibulo-ocular reflex in monkeys, *Trends Neurosci.*, **11**, 147–152 (1988).

[8]   Young, L. R., and Stark, L. W., A discrete model for eye tracking movements, *IEEE Trans. Mil. Electron. MIL.*, **7**, 113–115 (1963).

•   •   •   •   •

CHAPTER 16

Gait and Stance Control System

Humans are bipedal creatures who rely on appendages known as legs as the primary means of locomotion. The human legs are the lower limb of the body. They extend from the hip to the ankle, encompassing the knee and everything in between. The legs are simple instruments, and yet the complex coordination among each component in it is able to generate the required force and motion to move the body forward. This movement is called gaiting. Gaiting is defined as the manner in which locomotion is achieved by using the human limbs. Some common forms of human gaiting are walking, running, crawling, and hopping. Each of them requires varying starting orientations, stopping motions, changes in speed, alternations in direction, and modifications for changes in slope [1].

The motivation to understand the processes of human gaiting are beginning to take root in the medical world. The need for artificial legs by amputees to experience a normal daily life has pushed early medical pioneers to build various models for the legs. However, these models were too primitive to account for the complicated nature of gaiting. Today, rapid developments in biomedical research have provided the resources to finally make better advances in discovering the underlying mechanics of gaiting. Apart from restoring amputees, understanding human leg motions can provide a standard or a reference to compare normal and abnormal gaiting that is a result from pain, paralysis, tissue damage, or other motor control dysfunction. This would perhaps provide better efficiency in detecting, and possibly curing, those problems.

The modeling of the human gait focuses on three components of the leg. The hips, knee, and ankle are the primary areas of research. Although all these three parts work together to provide motion, there have been only isolated studies for each. Many articles and journals provide insight on only one of the three components. This could be because the coordination of all three is relatively complex and would require an excess of resources to combine them. One known application for all three would be the construction of robots to mimic human gaiting. Scientists in Japan had produced a mechanical automaton called the ASIMO (Advanced Step in Mobility). The ASIMO is a robot built specializing in emulating human gaiting. Much of the work on the robot focuses on balance and coordination. The motion of the legs can be easily emulated, but without coordination and feedback among the hips, knee, and ankle, the robot continues to fumble in its initial design [5].

The leg structure is commonly characterized by a set of interconnecting pendulum systems. In Figure 16.1, it can be observed that the first pendulum system begins from the hip to the knee. The swinging angle from the hip will determine the displacement of the knee. A second pendulum system connects the knee to the ankle. Knee movements provide for the position on how the ankle lands. During gaiting, two different situations arise in sequence: the statically stable supported phase, in which the whole body is kept aloft by both legs simultaneously; and the statically unstable support phase, in which only one foot is in contact with the ground while the other is being transferred from the back to the front. In a single walking cycle, the kinematic structure of locomotion changes from an open to a closed kinematic chain [4].

The coordination of the interconnecting pendulum assumptions can be further observed the simple gaiting (walking) process in Figure 16.2. The human walking is a process of locomotion in which the erect, moving body is supported by first one leg and then the other; as the moving body passes over the supporting leg, the other leg swings forward in preparation for its next support phase. One foot of the other is always on the ground, and during that period when the support of the body is transferred from the trailing to the leading leg there is a brief period when both feet are on the ground [2].

As walking speed increases, the periods of double support become smaller fractions of the walking cycle until, eventually, as a person starts to run, they disappear altogether and are replaced by brief periods when neither food is on the ground. The cyclic alternations of the support function of each leg and the existence of a transfer period when both feet are on the ground are essential features of the locomotion process known as walking.

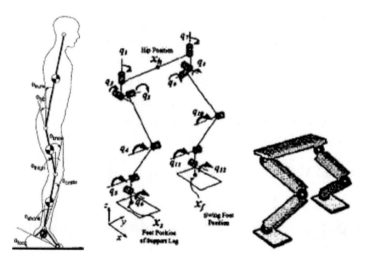

FIGURE 16.1: The leg system of the human body.

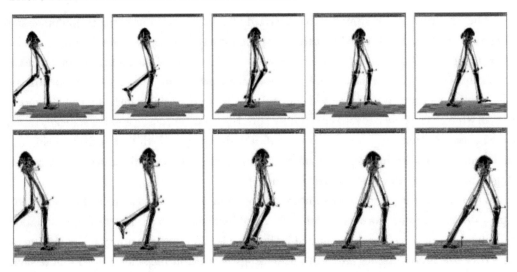

FIGURE 16.2: The phases of a normal gaiting process (breeze walking).

In attempting to understand the processes of gaiting and to combine the models for the hips, knee, and ankle together, Figure 16.3 displays the logic of a model. The pelvis, or the hips, is set as the starting point with the ankle as the ending point. The hips would receive an input signal from the central nervous system (CNS) or the peripheral nervous system, which is dependent on the type of action intended. Mainly for our model, input would come from the CNS; there will be no reflex reactions. The hip would influence the knee, and in turn the knee would determine the reaction of the ankle. The system of communication between the three components is in series with feedbacks coming out from each of their respective outputs. The final output would be a change in displacement or position [3].

Several assumptions were made for the modeling. The whole gating process is assumed to take place from the hips and down. The process for moving is totally isolated to just the coordination between the hips, knee, and ankle. There will also be no additional loads applied to the legs other than the upper body. The legs will just be carrying the mass of the entire human body and will not be subjected to external forces. The walking surface for the leg model is assumed to be a flat surface with adequate friction coefficient for motion. Because the model focuses on the lower body,

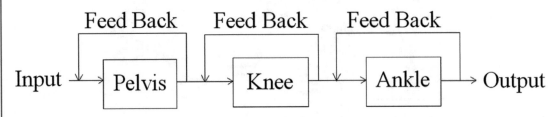

FIGURE 16.3: The block diagram of the mechanical gaiting coordination system.

influence from the vestibular feedback input and visual perception are ignored. The legs are made to act independently of their input. The arms will not be involved in balancing the body while in motion. Also, the legs are assumed to be in two-dimensional motions.

The model is first broken into the common three components; the hips, knee, and ankle. Each of these segments will be individually observed. Transfer functions were obtained for each component. Laboratory Virtual Instrumentation Engineering Workbench (LabVIEW) virtual instruments (VIs) were also constructed to simulate the transfer function's responses, and then Bode plots, root locus, and Nyquist plots were developed for each transfer function. Finally, all segments were combined together to form an overall system that represent the entire leg. Again, LabVIEW was used to simulate the entire leg's response.

16.1 THE HIP

There are not many models that describe the human hip; hence, the mechanical model use in the LabVIEW program is shown in Figure 16.4. This part of the leg is modeled as a thighbone with an actuator behind the bone. An analytic relation between the force provided by the actuator and the torque needed to move the articulation has been found. In particular, the segment k represents the arm of the hip joint torque for the model. The reaction force due to gravity is Mg, where M is the mass of the leg, and g is the gravitational acceleration due to gravity. The length of the thighbone is L, and from the diagram, k is found to be a proportion of L, in this case 0.7. The angle ω is between the actuator and the arm of the hip joint torque, and is treated as constant for this model. From observation, is it assumed to be 10° or 0.175 rad.

$$P(s) = \frac{MgL}{k\sin(\omega)}\left(\frac{s}{s^2 + 1}\right)$$

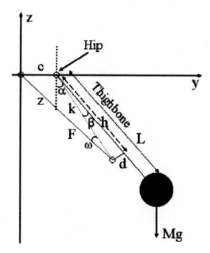

FIGURE 16.4: Mechanical model for the hips.

where

L = length of thighbone = 48 cm = 0.48 m

M = mass of leg (men, 19.5–25.2 kg; women, 11.7–16.6 kg)

g = acceleration due to gravity (9.80 m/s^2)

k = arm of the hip joint torque, estimated from diagram to be 70% of L = 33.6 cm

ω = estimated to be 10° (=0.175 rad), low angle so sin(ω): ω = 0.175

Figures 16.5 and 16.6 show the LabVIEW front panel view and block diagram, which contain the control VIs used in developing the LabVIEW hip program.

16.2   THE KNEE

To model the human knee, the tendon can be treated as a small torsional spring damper with inertia (J), stiffness (K), and damping (B) (Figure 16.7) Engineers try to estimate these parameters through experimental data from real human knees. When the tendon is excited, a signal is sent through the

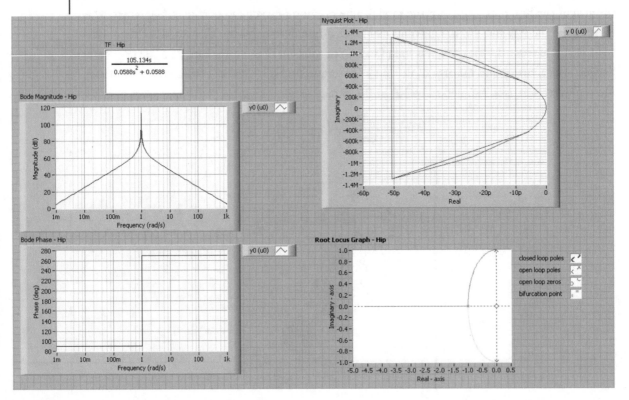

FIGURE 16.5: Hip VI front panel.

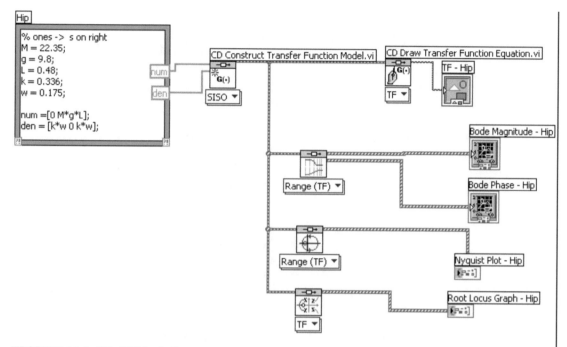

FIGURE 16.6: Hip VI block diagram.

nervous system to the spinal cord reporting a structural change (that is tendon length). The nervous system then sends a signal back to the tendon to produce a reflex. There are receptors on the muscle called spindles, which have their own dynamics and are shown in the model as a transfer function in the feedback path. These spindles are modeled as a spring (K_{pe}) and damper (B_{pe}) in parallel, and then with the pair in series with another spring (K_{se}).

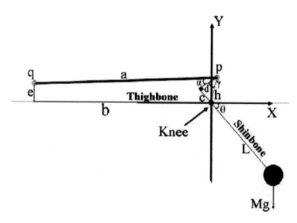

FIGURE 16.7: Mechanical model of the knee.

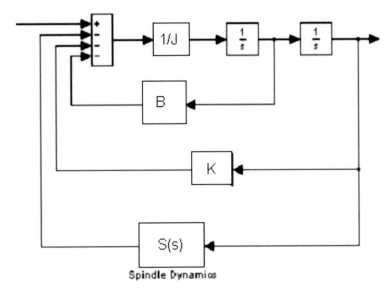

FIGURE 16.8: Block diagram of knee.

Figure 16.8 is a commonly used block diagram to reason out the process of the knee, where

$$S(s) = \frac{1}{K_{se}(B_{pe}s + K_{pe})}$$

For the feedback system,

$$G(s) = \frac{G(s)}{1 + G(s)H(s)}$$

Therefore,

$$G_1(s) = \frac{\dfrac{1}{Js}}{1 + \dfrac{B}{Js}} = \frac{1}{Js + B}$$

$$G_2(s) = \frac{\dfrac{1}{Js^2 + Bs}}{1 + \dfrac{K}{Js^2 + Bs}} = \frac{1}{Js^2 + Bs + K}$$

where $B = 2.4382$, $J = 0.19033$, $K = 42.361$, $K_{pe} = 4.7627$, $B_{pe} = 0.96703$, and $K_{se} = 0.10774$.

Figures 16.9 and 16.10 show the LabVIEW front panel view and block diagram, which contain the control VIs used in developing the LabVIEW knee program.

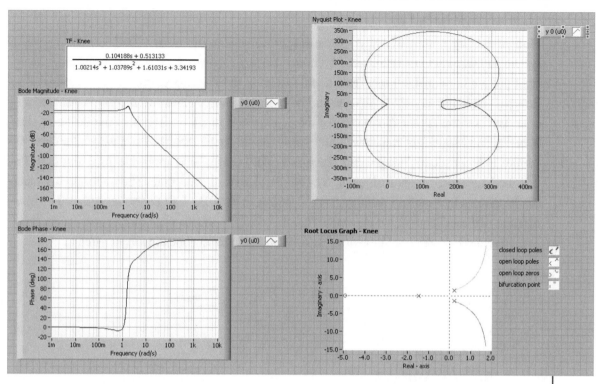

FIGURE 16.9: Front panel of knee.

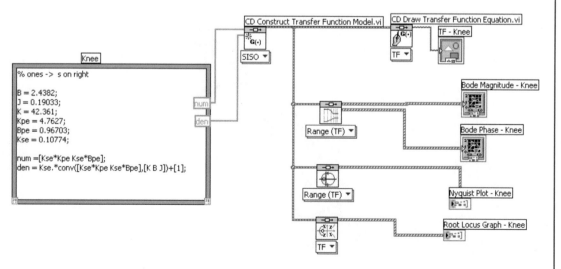

FIGURE 16.10: Block diagram of knee.

16.3   THE ANKLE

Human standing posture in sagittal plane, as approximated to an inverted pendulum, is an unstable system and requires to be stabilized. Several sensory organs seem to be used for stabilizing the upright posture in normal subject. The skeletal system of the standing posture is approximated as an inverted pendulum in sagittal plane as shown in Figure 16.11. Knee and hip joints were fixed by brace and ignored. Although the sway of the pendulum is small, the inertia of the skeletal system, viscosity, and elasticity of the muscle, and gravity effect can be described as a two-order delay dynamics. A proportional, integral, and derivative (PID) controller is applied to $M(s)$ and $G(s)$. The muscle characteristic is approximated to one-order delay and dead time. The feedback parameters can be decided by a model matching method similar to the PID joint controller design. The block diagram of the ankle process is shown in Figure 16.12.

From the dynamics of human ankle stiffness, the transfer function $G(s)$ is given as

$$G(s) = \frac{\theta(s)}{T(s)} = \frac{1}{Js^2 + Bs + K - \dfrac{mgl}{2}}$$

where

$\theta(s)$ = ankle joint angle

$T(s)$ = moment of ankle joint torque; at 20 Nm external torque

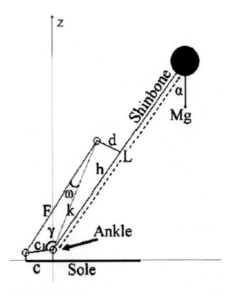

FIGURE 16.11: Mechanical model of ankle.

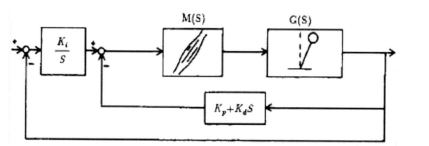

Where:

$K_i = 15$,

$K_p = 15$, and

$K_d = 23$.

FIGURE 16.12: Block diagram of the ankle process.

J = moment of ankle joint inertia (0.008 N m s^2/rad)

B = moment of ankle joint viscosity (10 N m s/rad)

K = moment of ankle joint elasticity (240 N m/rad)

m = mass of human (80 kg)

l = height of human (1.7 m)

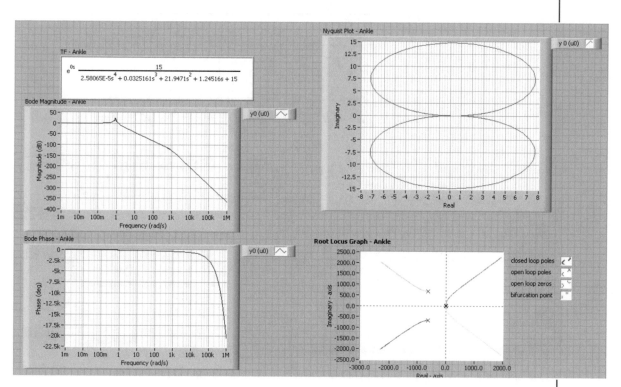

FIGURE 16.13: Front panel of ankle.

Then, with 0 external torque, $M(s)$ is calculated as

$$M(s) = \frac{K_m e^{-Ds}}{1 + \tau s}$$

where

K_m = gain of muscle (31 N m/rad)

τ = time constant of muscle (=0.1 s for f = 10 Hz)

D = dead time of muscle (~350 μs)

$$G_1(s) = M(s) G(s) = \left(\frac{1}{Js^2 + Bs + K - \dfrac{mgl}{2}} \right) \left(\frac{K_m e^{-Ds}}{1 + s} \right)$$

$$G_{1H}(s) = \frac{\left(\dfrac{1}{Js^2 + Bs + K - \dfrac{mgl}{2}} \right) \left(\dfrac{K_m e^{-Ds}}{1 + \tau s} \right)}{1 + \left(\dfrac{1}{Js^2 + Bs + K - \dfrac{mgl}{2}} \right) \left(\dfrac{K_m e^{-Ds}}{1 + \tau s} \right) (K_p + K_d s)}$$

$$= \frac{1}{\left(Js^2 + Bs + K - \dfrac{mgl}{2} \right) \left(\dfrac{1 + \tau s}{K_m e^{-Ds}} \right) + (K_p + K_d s)}$$

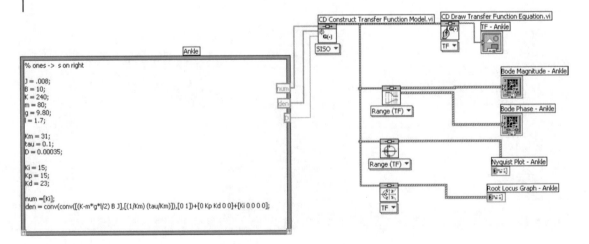

FIGURE 16.14: Block diagram of ankle.

$$G_2(s) = G_{1\mathrm{H}}(s)\left(\frac{K_\mathrm{i}}{s}\right) = \cfrac{K_\mathrm{i}}{\left(Js^2 + Bs + K - \dfrac{mgl}{2}\right)\left(\dfrac{1+\tau s}{K_\mathrm{m}e^{-Ds}}\right)s + \left(K_\mathrm{p}s + K_\mathrm{d}s^2\right)}$$

$$G_x(s) = \cfrac{\cfrac{K_\mathrm{i}}{\left(Js^2 + Bs + K - \dfrac{mgl}{2}\right)\left(\dfrac{1+\tau s}{K_\mathrm{m}e^{-Ds}}\right)s + \left(K_\mathrm{p}s + K_\mathrm{d}s^2\right)}}{1 + \cfrac{K_\mathrm{i}}{\left(Js^2 + Bs + K - \dfrac{mgl}{2}\right)\left(\dfrac{1+\tau s}{K_\mathrm{m}e^{-Ds}}\right)s + \left(K_\mathrm{p}s + K_\mathrm{d}s^2\right)}}$$

$$= \cfrac{K_\mathrm{i}}{\left(Js^2 + Bs + K - \dfrac{mgl}{2}\right)\left(\dfrac{1+\tau s}{K_\mathrm{m}e^{-Ds}}\right)s + \left(K_\mathrm{p}s + K_\mathrm{d}s^2\right) + K_\mathrm{i}}$$

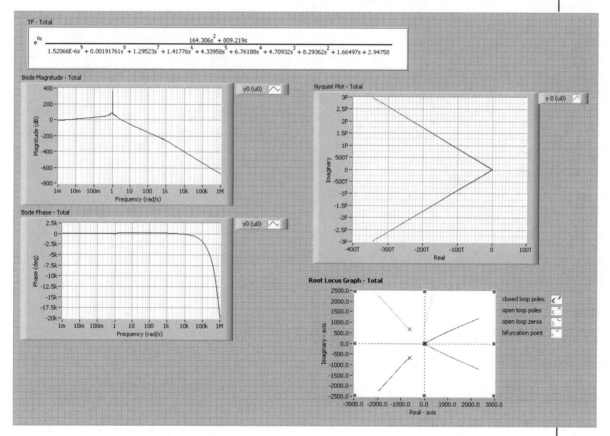

FIGURE 16.15: Front panel of overall.

Figures 16.13 and 16.14 show the LabVIEW front panel view and block diagram, which contain the control VIs used in developing the LabVIEW ankle program.

16.4   OVERALL SYSTEM

Figures 16.15 and 16.16 are the modeling of the overall system in LabVIEW. All the transfer functions obtained are assumed to work in series. The response of the system is simulated and displayed.

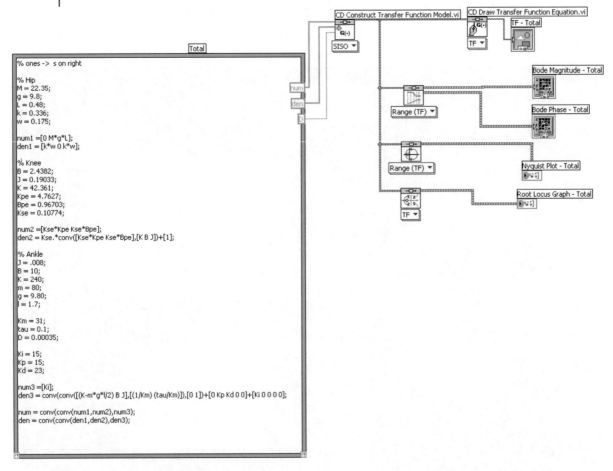

FIGURE 16.16: Block diagram of overall.

REFERENCES

[1] Rose, J., and Gamble, J. G., *Human Walking*, 3rd ed., Lippincott Williams & Wilkins, Philadelphia (2006).

[2] Giannini, S., Catani, F., Benedetti, M. G., and Leardini, A., *Gait Analysis: Methodologies and Clinical Applications*, 1st ed., IOS P, Amsterdam, Netherlands (1994).

[3] Vaughan, C. L., Davis, B. L., and O'Connor, J. C., *Dynamics of Human Gait*, Human Kinetics, Campaign, IL (1992).

[4] Muscato, G., and Spampinato, G., Kinematical model and control architecture for a human inspired five DOF robotic leg, *Mechatronics*, 17, 45–63 (2007), <http://www.sciencedirect.com/science/article/B6V43-4KVXPKF-1/1/090b0a36e51f67445c5a4ceaad969241>, accessed 2 Apr. 2008.

[5] Kim, J.-Y., Park, I. W., and Oh, J.-H., Experimental realization of dynamic walking of the biped, *Adv. Robotics*, 20, 707–736 (2006).

CHAPTER 17

Respiratory Control System

17.1 PULMONARY PHYSIOLOGY

The primary purpose of the respiratory system is to obtain oxygen for the body to use in metabolism, and to eliminate the waste product, carbon dioxide. External respiration is the exchange of gases between the atmosphere and the body. This includes ventilation, gas exchange at the lung and the cells, and the transport of gases in the blood. External respiration is controlled by three primary factors: CO_2, O_2, and pH. A change in any of these factors can result in a range of responses in respiration.

17.2 BASICS

As stated, the three main chemical factors responsible for respiratory control are carbon dioxide (CO_2), oxygen (O_2), and pH. Each of these factors has a desired value at various points in the body. These values are shown in Table 17.1. Ventilation control occurs in response to deviations from these values.

The response to an increase in partial pressure of oxygen is a decrease in ventilation, whereas a decrease in pO_2 will result in an increase in ventilation. The respiratory system responds in the opposite fashion for partial pressure of carbon dioxide and pH, which are related through the Henderson–Hasselbalch equation. For example, an increase in arterial pCO_2 results in increased rate and depth of ventilation, thus increasing alveolar ventilation and removing CO_2 from the blood.

17.3 METHOD OF VENTILATION CONTROL

Ventilation is considered to be the movement of air in and out of the lungs. This process is controlled by the central nervous system (CNS). Respiratory control is performed by the central pattern generator (CPG), which is a network of neurons in the brain stem. The CPG functions automatically, creating rhythmic cycles of neuronal fire for inspiration and expiration. This cycle is continuously influenced by sensory input from the chemoreceptors. The three primary factors for respiratory control are CO_2, O_2, and pH.

The chemoreceptors that monitor these factors are located peripherally and locally in the CNS. Peripheral chemoreceptors are located in the carotid and aortic bodies. They sense fluctua-

TABLE 17.1: Normal blood values in the pulmonary system		
FACTOR	**ARTERIAL**	**VENOUS**
pO_2	100 mm Hg	40 mm Hg
pCO_2	40 mm Hg	46 mm Hg
pH	7.4	7.37

tions in CO_2, O_2, and pH, and send the information to the CPG to regulate ventilation. The peripheral chemoreceptors are more responsive to changes in pO_2 than to changes in plasma pH and pCO_2. However, under most circumstances, oxygen is not important in the regulation of ventilation because arterial pO_2 must drop to 60 mm Hg before ventilation is stimulated. It would take an unusual physiological condition to cause this response.

The central chemoreceptors are located in the brain, on the ventral surface of the medulla. The central chemoreceptors respond to changes in the concentration of carbon dioxide in the cerebral spinal fluid (CSF). Stating that the central chemoreceptors monitor CO_2 is misleading, because the central chemoreceptors actually respond to pH changes in the CSF caused by the conversion of CO_2 into bicarbonate ion and H^+ as CO_2 diffuses across the blood–brain barrier into the CSF. These receptors are responsible for setting the respiratory pace by providing a constant input into the CPG.

Once the CPG receives a new setpoint signal, it sets a new ventilation pattern and respiration occurs. The inspiratory neurons that control the somatic motor neurons to the diaphragm are located in the medullary dorsal respiratory group. The ventral respiratory group controls muscles that are used for active respiration.

During quiet respiration, the inspiratory neurons gradually increase stimulation of the inspiratory muscles for 2 s, acting like a ramping signal. The process starts off with a few neurons firing, but the firing of these neurons recruits more and more inspiratory neurons to fire, acting like a positive feedback loop. As a result, the rib cage expands while the diaphragm contracts.

After 2 s of the inspiratory neuron ramping, they immediately shut off allowing the muscles to relax. The next 3 s consist of passive respiration, in which only minute amounts of neuronal activity exist. Passive respiration occurs in this manner because of the elastic recoil of the inspiratory muscles and the lung tissue.

Forced breathing occurs via a slightly different process. There are additional inspiratory muscles that are required because of the increased activity of the inspiratory neurons. The additional muscles used for forced inspiration are the sternocleidomastoids. Active expiration then occurs through stimulation provided by the ventral respiratory group, which incorporates the internal intercostals and the abdominal muscles.

There are protective reflexes that guard the lungs from irritation and overexpansion. Bronchoconstriction, which is mediated through parasympathetic neurons, protects the lungs from irritants. The Hering–Breuer inflation reflex is responsible for preventing overexpansion. If tidal volume exceeds 1 L, the stretch receptors in the lung send a signal to the brain to stop inspiration.

17.4   GAS LAWS

Air flow follows the laws of diffusion. Air flow is directed from areas of higher pressure to areas of lower pressure. For instance, as oxygen is transported throughout the body, the flow is from areas of higher partial pressure to areas of lower partial pressure. Arterial pO_2 is always higher than venous pO_2, which explains why air flow constantly moves throughout the system.

Boyle's law describes the pressure–volume relationships of gases, as shown by Equation (17.1).

$$P_1 V_1 = P_2 V_2 \qquad (17.1)$$

Boyle's law explains how in respiration changes in the volume of the chest cavity during ventilation create different pressure gradients, which are responsible for inducing air flow.

The volume change is created by a pump that, for the respiratory system, consists of the muscles of the thoracic cage as well as the diaphragm. Pressures are measured with respect to the air spaces of the lungs (alveolar pressure) or the intrapleural pressure. For instance, during expiration, when chest volume decreases, alveolar pressure increases, and air flows out of the respiratory system. This is an example of bulk flow, because the entire mixture of gases is being transported.

The efficiency of breathing is estimated through the calculation of the total pulmonary ventilation (TPV). TPV is the volume of air moved into and out the lungs each minute. The calculation is stated as "Total pulmonary volume is equal to the product of the ventilation rate times the tidal volume" [1].

Although TPV is a good indicator of how much air moves in and out of the respiratory tract, it does not indicate the amount of fresh air reaching the alveolar exchange surface. Therefore, true efficiency can be calculated with the calculation of alveolar ventilation as follows [1]:

$$\text{Alveolar ventilation} = \text{Ventilation rate} \times (\text{Tidal volume} - \text{Dead space}) \qquad (17.2)$$

Alveolar ventilation can be considerably affected by the rate and depth of breathing. Breathing as quickly and deeply as possible is called maximum voluntary ventilation. This is an abnormal case, however. The normal values for TPV and AV are shown in Table 17.2.

TABLE 17.2: Normal ventilation values in the pulmonary system	
Total pulmonary ventilation	6 L/min
Total alveolar ventilation	4.2 L/min
Respiration rate	12–20 bpm
Tidal volume	500 mL

17.5   GAS EXCHANGE AT THE ALVEOLI

The first step in external respiration is the movement of oxygen from the atmosphere to the alveolar exchange surface in the lungs. Gas exchange then occurs across the alveolar–capillary interface. Perfusion must be high enough past the alveoli in order to pick up the available oxygen. If blood pressure falls below a certain point, the capillaries close off. This results in a diversion of blood flow to another capillary bed that is seeing a higher blood pressure.

Throughout this process, local regulators are attempting to match blood and air flow through regulation of the arteriole and bronchiole diameter. The bronchiole diameter is mediated by levels of carbon dioxide in the exhaled air. If there is an increase in the partial pressure of exhaled carbon dioxide, bronchodilation occurs; however, the converse is true for a pCO_2 decrease.

Resistance of the pulmonary arterioles to blood flow is regulated by the oxygen content of the interstitial fluid. When the pO_2 is increased, the result is arteriole dilation. The converse is true for decreased pO_2.

17.6   GAS EXCHANGE IN THE LUNGS AND TISSUES

Gas laws state that individual gases flow from regions of higher partial pressure to regions of lower partial pressure, which is the governing rule for gas exchange in the lungs and tissues. Note in Table 17.1 that oxygen moves down its gradient from the alveoli into the capillaries. This diffusion reaches equilibrium and thus pO_2 of arterial blood leaving the lungs is the same as the alveoli. When this blood reaches the capillaries of the tissue, the gradient is reversed. Cells are always using O_2 for oxidative phosphorylation; therefore, oxygen moves down its gradient as it leaves the plasma and goes into the cells. Again, diffusion reaches equilibrium, which results in venous blood having the same pO_2 rate as the cells it has just passed.

The opposite process occurs with respect to carbon dioxide. pCO_2 is higher in the tissues than systemic capillary pCO_2 because cells are creating CO_2 as a waste product of metabolism. The resulting gradient brings CO_2 out of cells into the capillaries. Diffusion occurs and venous blood increases in pCO_2.

17.7   GAS EXCHANGE IN THE BLOOD

Oxygen can be transported in the blood in two different methods. The most common method (98%) of oxygen transport occurs when oxygen binds to hemoglobin in red blood cells. The other method is through dissolution in the plasma. However, because of oxygen's slight solubility, there is a great reliance on the hemoglobin method.

The oxygen–hemoglobin binding obeys the law of mass action, which states for this application: if oxygen concentration increases, the binding reaction shifts so that more oxygen binds to hemoglobin. If O_2 decreases, the reaction shifts so that hemoglobin releases some of its bound oxygen. Therefore, the amount of oxygen bound to hemoglobin depends primarily on the pO_2 of the plasma surrounding the red blood cells. With increased metabolic activity, pO_2 decreases, resulting in a release of oxygen from the hemoglobin.

Carbon dioxide is transported through the blood by several different means. Carbon dioxide is more soluble in plasma compared to oxygen, but too much is produced to be completely dissolved in the plasma. Nearly a quarter of the CO_2 binds to hemoglobin, and the rest are converted to bicarbonate ion. CO_2 removal from the body is extremely important to prevent acidosis and depression of the CNS function.

17.8   CONCEPTUAL MODEL

In the development of the block diagram for the conceptual model (Figure 17.1), the following assumptions were made:

1. The model will be a three-compartment model consisting of lung, brain, and bodily tissues. The pO_2/pCO_2 in the brain is monitored by the central chemoreceptors and the pO_2/pCO_2 in the body is monitored by the peripheral chemoreceptors in the carotid and aortic bodies. The pO_2/pCO_2 in the lung tissue is monitored by capillaries in the alveoli and determines local lung perfusion.
2. pH is related to pCO_2 via the Henderson–Hasselbalch equation [2,3] and can be eliminated as an input.
3. There is no conscious input into the respiratory system.

The model consists of the following components:

1. The CPG, which controls the activity of the efferent neurons that activate the inspiratory and expiratory muscles.

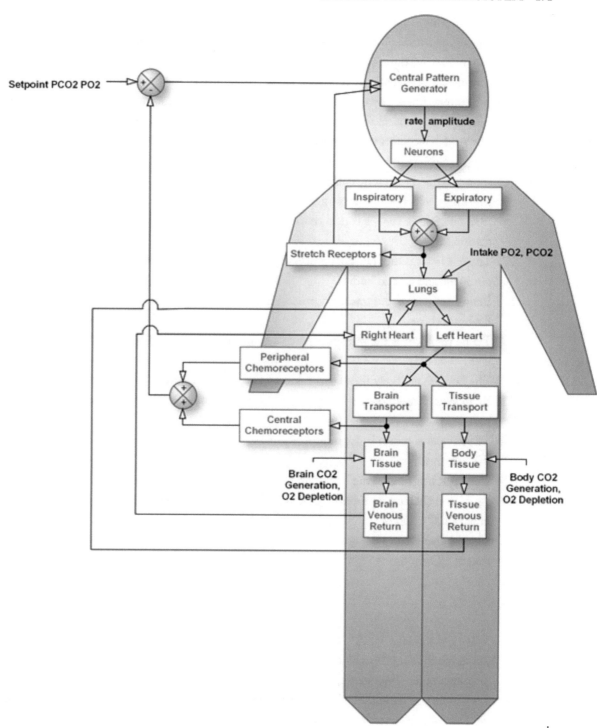

FIGURE 17.1: Preliminary block diagram for the respiratory system.

2. The neurons responding to inputs from the CPG. The number of active inspiratory neurons as a function of time is a periodic ramp signal in quiet breathing. Expiratory neurons remain inactive during quiet breathing.

3. The inspiratory muscles that include the diaphragm, external intercostals, scalene, and sternocleidomastoids. The diaphragm contributes most to changes in lung volume.

4. The expiratory muscles that include the internal intercostals and abdominal muscles.

5. The stretch receptors that prevent overexpansion of the lungs. Stretch receptors in the lung trigger the brain to terminate inspiration when the tidal volume exceeds 1 L. This phenomenon is termed the Herring–Breuer inflation reflex.

6. The lungs, which is the site of gas exchange. The CO_2 and O_2 concentrations in the lungs are a function of venous concentrations, intake air concentration, lung perfusion, and ventilation rate.

7. The left heart is the transport process from the lungs to the peripheral chemoreceptors in the aortic and carotid bodies.

8. Brain transport is the transport process from the main arteries to the arterioles that feed the brain tissue.

9. Tissue transport is the transport process from the main arteries to the arterioles that feed bodily tissues.

10. The brain tissue/body tissue that takes up O_2 and contributes CO_2.

11. Brain venous return is the transport process that returns blood from the brain to the right heart.

12. Tissue venous return is the transport process that returns blood from the bodily tissues to the right heart.

13. The right heart mixes the brain and tissue venous return, and outputs blood to the lung.

14. Peripheral chemoreceptors are located in the carotid and aortic arteries, and sense changes in pO_2, CO_2, and pH.

15. Central chemoreceptors are located in the brain and respond to changes in the concentration of CO_2 in the cerebrospinal fluid.

17.9   MATHEMATICAL MODEL

Figure 17.2 shows the simplified block diagram for the mathematical model used in the Laboratory Virtual Instrumentation Engineering Workbench (LabVIEW) Control program focusing on the impact of pCO_2 on ventilation. In addition to the previous assumptions, the following assumptions were made. The variables used in the model are listed in Table 17.3.

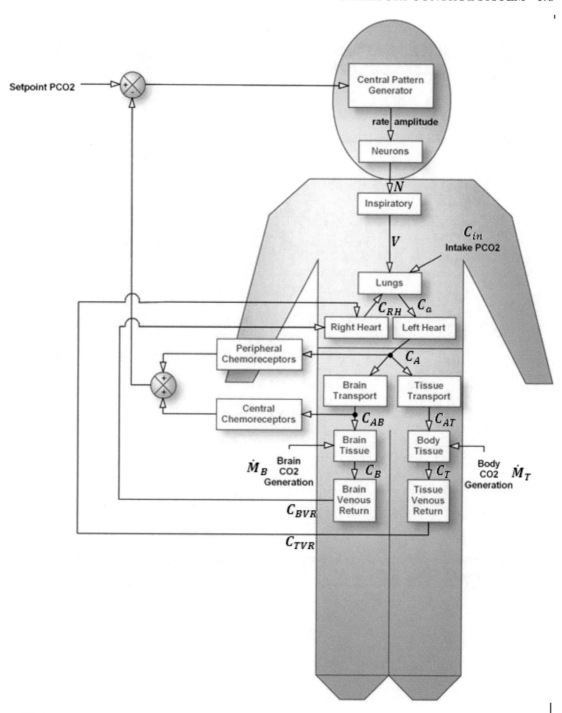

FIGURE 17.2: Simplified block diagram for mathematical model.

TABLE 17.3: List of variables	
PRIMARY VARIABLES	
N	Number of active inspiratory neurons
V	Tidal volume of the lungs
C_a	Concentration of CO_2 in the alveoli
C_{RH}	Concentration of CO_2 in blood coming from right heart
C_{in}	Concentration of CO_2 in inspired air
C_A	Concentration of CO_2 in the major arteries and peripheral chemoreceptors
C_{AT}	Concentration of CO_2 in the arterioles feeding the body tissue
C_{AB}	Concentration of CO_2 in the arterioles feeding the brain tissue
C_T	Concentration of CO_2 in body tissue compartment
C_B	Concentration of CO_2 in brain tissue compartment
$\dot{M}_T$	Rate of CO_2 generation in body tissue
$\dot{M}_B$	Rate of CO_2 generation in brain tissue
C_{TVR}	Concentration of CO_2 in the venous return from the body tissue compartment
C_{BVR}	Concentration of CO_2 in the venous return from the brain tissue compartment
PHYSIOLOGICAL PARAMETERS	
k_L	Elastance of diaphragm system
m_L	Inertance of diaphragm system
B_L	Damping of diaphragm system

TABLE 17.3: (*continued*)	
PHYSIOLOGICAL PARAMETERS	
K_f	Measure of muscle generated force and volume expansion as a function of number of active neurons
$\dot{Q}$	Cardiac output (perfusion rate)
V_T	Volume of the body tissue compartment
$\dot{Q}_T$	Body tissue perfusion rate
V_B	Volume of the brain tissue compartment
$\dot{Q}_B$	Brain tissue perfusion rate
V_{RH}	Volume of the right heart

17.10 ADDITIONAL ASSUMPTIONS

1. The circulation times between compartments are finite and constant.
2. For quiet respiration, expiration is passive and expiratory muscles can be excluded from the mathematical model.
3. Under normal circumstances, the arterial pO_2 has little effect on ventilation control. pO_2 must be lower than 60 mm Hg before this input can exert a significant impact on ventilation. Thus, the mathematical model will focus on the treatment of pCO_2.
4. The stretch receptor reflex is only activated during strenuous exercise and can be ignored under normal respiratory conditions.
5. The concentration of CO_2 in each compartment is uniform.
6. All of the CO_2 generated by the body and brain tissue compartments is transferred to the blood.

17.11 DERIVATION OF EQUATIONS
17.11.1 Inspiratory Muscles

Inspiratory muscles translate neuronal signals into lung volume. The relationship between the number of active inspiratory neurons and the lung volume will be modeled using the mechanical model given in Figure 17.3. In this model, the diaphragm is assumed to undergo a one-dimensional linear

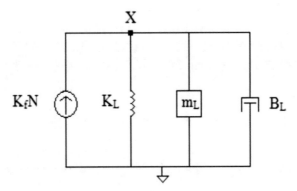

FIGURE 17.3: Mechanical circuit diagram for diaphragm displacement.

displacement. The force generated by the muscle is a function of the number of active neurons. The diaphragm, which is the major inspiratory muscle, can be modeled with a certain elastance K_L, a mass m_L, and a damping constant B_L. The displacement of the diaphragm, X, can be used as an indicator of lung volume.

The diaphragm displacement can be expressed in terms of a differential equation as shown in Equation (17.4).

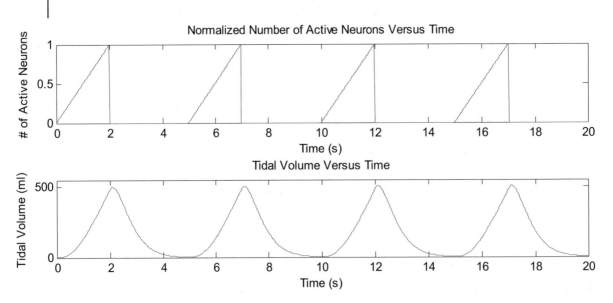

FIGURE 17.4: Matlab simulation of inspiratory muscle transfer function.

$$K_f N = m_L D^2 x + B_L Dx + Kx \qquad (17.3)$$

If volume change (V) is assumed to be a linear function of the displacement (x), the new constant of proportionality can be incorporated into K_f, which now has units of $g \cdot cm^3 \cdot s^{-2}$. Note also that m_L is expressed in grams, B_L in $g \cdot s^{-1}$, and K_L in $g \cdot s^{-2}$. After these changes, the Laplace transform of the differential equation can be used to find the function given in [1]

$$V(s) = \frac{K_f}{m_L s^2 + B_L s + K_L} N(s) \qquad (17.4)$$

To find reasonable values for the parameters in Equation (17.4), Matlab was used to simulate the output to a periodic ramp signal input. The ramp input was normalized so that its amplitude was 1. In accordance with Silverthorn [4], the inspiratory phase was set to 2 s and the expiratory phase was set to 3 s. Results ($m_L = 600$ g, $B_L = 2800$ $g \cdot s^{-1}$, and $K_L = 3500$ $g \cdot s^{-2}$, and $K_f = 2.8 \times 10^6$ $g \cdot cm^3 \cdot s^{-2}$) are shown in Figure 17.4.

17.11.2 Lungs

In the lungs, the exchange of CO_2 is based on the relative concentrations of CO_2 inside the alveoli, venous blood, and gas input. The relationship between the amount of CO_2 in the alveoli and these variables can be expressed in the following differential equation

$$\frac{d(VC_a)}{dt} = \begin{cases} k_1 \dot{Q}(C_{RH} - C_a) + \dot{V}C_{in}; \ \dot{V} \geq 0 \\ k_1 \dot{Q}(C_{RH} - C_a) + \dot{V}C_a; \ \dot{V} < 0 \end{cases} \qquad (17.5)$$

where

V = lung volume
C_a = concentration of CO_2 in the alveoli
C_{RH} = concentration of CO_2 in the blood from the right heart
C_{in} = concentration of CO_2 in the incoming gas
$\dot{Q}$ = perfusion rate
$\dot{V}$ = ventilation rate (the time derivative of lung volume)
k_1 = constant that describes the efficiency of gas transport between the pulmonary capillaries and the alveoli [1]

Equation 5 essentially states that during inspiration, the time rate of change of the total amount of CO_2 in the arterioles is equal to the contribution from right heart perfusion, $k_1 \dot{Q}(C_{HR} - C_a)$ plus

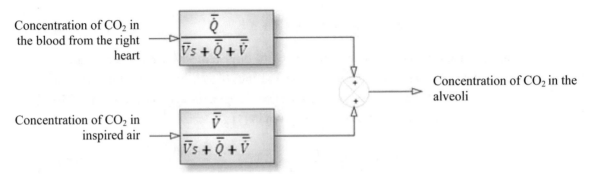

Concentration of CO_2 in the blood from the right heart

$$\frac{\bar{\bar{Q}}}{\bar{V}s + \bar{\bar{Q}} + \bar{\bar{V}}}$$

Concentration of CO_2 in inspired air

$$\frac{\bar{\bar{V}}}{\bar{V}s + \bar{\bar{Q}} + \bar{\bar{V}}}$$

Concentration of CO_2 in the alveoli

FIGURE 17.5: The block diagram for the alveolar CO_2 concentration.

the influx of inspired air ($\dot{V}C_{in}$). During expiration, the time rate of change of the total amount of CO_2 in the arterioles is equal to the contribution from right heart perfusion and the expiration of air currently in the arterioles ($\dot{V}C_a$).

Equation (17.5) is similar to the one given in Milhorn's model of the respiratory system. However, Milhorn assumed that k_1 was unity, distinguished between alveolar–arterial CO_2 concentration and the alveolar CO_2 concentration, and used average values for lung volume and ventilation rate. A slight modification of Milhorn's equation is given as [1]

$$\frac{\bar{V}(dC_a)}{dt} = \bar{\bar{Q}}(C_{HR} - C_a) + \bar{V}(C_{in} - C_a) \tag{17.6}$$

Taking the Laplace transform of Equation (17.6) and expanding the result produces Equation (17.7). The block diagram for the alveolar CO_2 concentration is shown in Figure 17.5.

$$\left(\bar{Vs} + \bar{\bar{Q}} + \bar{V}\right)C_a(s) = \bar{\bar{Q}}C_{RH}(s) = \bar{\bar{Q}}C_{RH}(s) + \bar{V}C_{in}(s) \tag{17.7}$$

17.11.3 Left Heart

The time for transport from the lungs to major arteries are assumed to be finite and constant. According to Topor et al. [5], the circulatory transport delay between the lungs and the carotid bodies is 7.8 s. Furthermore, mixing in the left ventricle can be modeled as a first-order low-pass filter with a time constant of 1.5 s [5]. Thus, assuming unity gain, the left heart can be modeled as shown below

$$C_A(s) = \frac{e^{-7.8s}}{\dfrac{s}{1.5} + 1}C_a(s) \tag{17.8}$$

17.11.4 Brain and Tissue Transport

According to Topor et al. [5], the circulatory transport delay between the lungs and the brain compartment is 11.4 s. Subtracting the transport delay from the lungs to the major arteries, this leads to a transport delay of 3.6 s from the major arteries to the brain. Thus, arterial transport to the brain compartment can be modeled as

$$C_{AB}(s) = e^{-3.6s} C_A(s) \qquad (17.9)$$

Topor et al. [5] also define the control value for the circulatory delay between the lungs and the tissue compartment to be 18.6 s. Subtracting the transport delay from the lungs to the major arteries, this leads to a transport delay of 10.8 s from the major arteries to the tissue compartment. Thus, arterial transport to the tissue compartment can be modeled as

$$C_{AT}(s) = e^{-10.8s} C_A(s) \qquad (17.10)$$

Note that the brain and tissue transport transfer functions above do not account for arterial mixing because these time constants could not be found in the literature.

17.11.5 Body Tissue

The amount of CO_2 in the tissue can be modeled with the following differential equation.

$$V_T \frac{dC_T}{dt} = \dot{M}_T + \dot{Q}_T (C_{AT} - C_T) \qquad (17.11)$$

where

    V_T = volume of the tissue compartment
    C_T = CO_2 concentration in the tissue
    $\dot{M}_T$ = rate of CO_2 generation in the tissues
    $\dot{Q}_T$ = tissue perfusion rate
    C_{AT} = concentration of CO_2 in the arterioles feeding into the tissue

Equation (17.11) is given by Milhorn ([1], p. 237).

The Laplace transform of equation is given by

$$C_T(s) = \frac{\dot{M}_T + \dot{Q}_T C_{AT}(s)}{V_T s + \dot{Q}_T} \qquad (17.12)$$

This equation can be decomposed into the block diagram shown in Figure 17.6.

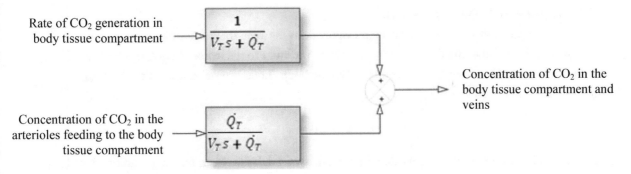

FIGURE 17.6: Decomposed block diagram for tissue and venous CO_2 concentration.

It is important to note at this point that although the dissolved CO_2 concentration only depends on the pCO_2, the total CO_2 concentration also depends on hemoglobin concentration, oxyhemoglobin saturation, and the dissociation constant of the CO_2–HCO_3^- system. According to Sun et al. [6], the total blood CO_2 concentration can be calculated using

$$C_{CO_2} = s \cdot p_{CO_2} \left(1 + 10^{pH - pK}\right) \left(1 - \frac{0.0289[Hb]}{3.352 - 0.456 \cdot S_{O_2}(8.142 - pH)}\right) \qquad (17.13)$$

where

   s = plasma solubility coefficient of CO_2
   pK = dissociation constant of the CO_2–HCO_3^- system
   S_{O_2} = oxyhemoglobin saturation
   $[Hb]$ = hemoglobin concentration

Thus, although the rate of CO_2 generation may be related to the total blood concentration of CO_2 as described by Milhorn's formula, its contribution to pCO_2 depends on many other physiological factors [1].

According to Topor et al. [5], the volume of the tissue compartment is 39.0 L. The tissue perfusion rate can be found by subtracting cerebral blood flow from total cardiac output; thus, $\dot{Q}_T$ can be calculated as 5.42 L/min [5].

17.11.6 Brain Tissue

The analysis applied to body tissues can be similarly applied to the tissues of the brain with some modifications in physiological parameters. The modified equatio n is shown in the following equation

$$C_B(s) = \frac{\dot{M}_B + \dot{Q}_B C_{AB}(s)}{V_B s + \dot{Q}_B} \tag{17.14}$$

The control parameters supplied by Topor et al. [5] indicated a cerebral blood flow, $\dot{Q}_B$, of 0.78 L/min and a brain compartment volume of 1.0 L.

17.11.7 Body and Brain Tissue Venous Return

Topor et al. [5] reported the circulatory delay between the brain compartment and the lungs to be 7.0 s, and the delay between the body tissue compartment to the lungs to be 33.6 s. Thus, the two venous return pathways can be modeled as shown in Equations (17.15) and (17.16). These transfer equations do not account for venous mixing.

$$C_{TVR}(s) = e^{-33.6s} C_T(s) \tag{17.15}$$

$$C_{BVR}(s) = e^{-7.0s} C_B(s) \tag{17.16}$$

17.11.8 Central and Peripheral Chemoreceptors

According to a study conducted by Ursino and Magosso [7] on the carotid chemoreceptor response, the response to CO_2 can be modeled as a linear comparison with a threshold level followed by a static gain constant and a single pole low-pass filter to simulate the "progressive attainment of a steady state level" (Ursino). Within the current framework of the mathematical model, comparison with a setpoint occurs after the chemoreceptor blocks. Thus, the chemoreceptors can be modeled as a low-pass filter with gains that determine their relative contribution to overall ventilation control.

Topor et al. [5] reported that the part of the inspiratory minute ventilation due to central chemoreceptor response is 5.62 L/min, and the part of the inspiratory minute ventilation due to peripheral chemoreceptor response is 1.54 L/min. Thus, the relative importance of the central and peripheral chemoreceptors can be estimated as 0.78 and 0.22, respectively. By using the time constant of 3.5 s provided by Ursino and Magosso for the carotid chemoreceptor for both the central and peripheral chemoreceptors and assuming that the contributions of the central and peripheral chemoreceptors are static, Equations (17.17) and (17.18) can be derived [7].

$$H_{central}(s) = \frac{0.78}{\frac{s}{3.5} + 1} \tag{17.17}$$

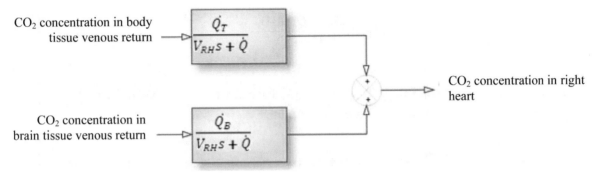

FIGURE 17.7: Decomposed block diagram for the right heart.

$$H_{peripheral}(s) = \frac{0.22}{\dfrac{s}{3.5} + 1} \qquad (17.18)$$

17.11.9 Right Heart

The right heart accepts and mixes the inputs from the body tissue and brain tissue, which can be modeled by using Equation (17.19).

$$V_{RH}\frac{dC_{RH}}{dt} = \dot{Q}_T (C_{TVR} - C_{RH}) + \dot{Q}_B (C_{BVR} - C_{RH}) \qquad (17.19)$$

where

V_{RH} = volume of the right heart

C_{TVR} = concentration of CO_2 in the venous return from body tissues

C_{BVR} = concentration of CO_2 in the venous return from brain tissues.

The Laplace transform of Equation (17.19) can be found in Equation (17.20).

$$\left(V_{RH}s + \dot{Q}\right) C_{RH}(s) = \dot{Q}_T C_{TVR}(s) + \dot{Q}_B C_{BVR}(s) \qquad (17.20)$$

This can be decomposed into the block diagram shown in Figure 17.7.

17.12   LABVIEW SIMULATIONS

LabVIEW was used to simulate the transfer functions for the inspiratory muscles, Milhorn's differential equation for the lung, the left heart, and chemoreceptors [1]. The block diagram for the inspiratory muscle simulation is shown in Figure 17.8. The front panel display with simulation

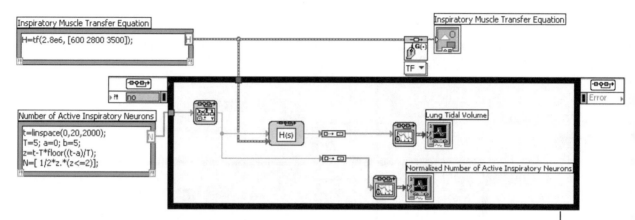

FIGURE 17.8: Block diagram for inspiratory muscle simulation.

results is shown in Figure 17.9. The mechanical constants for the diaphragm model were set so that the lung completely relaxed during expiration and reached a tidal volume of 500 mL.

The block diagram for the LabVIEW simulation of Milhorn's differential equation for the lungs is shown in Figure 17.10 [1]. The right heart blood CO_2 concentration was simulated using a sine wave with a mean of 26 mM and an amplitude of 1 mM. The average perfusion rate (cardiac output) was set as 100 mL/s; average lung tidal volume, 160 mL; average ventilation rate, 60 mL/s; and incoming CO_2 concentration, 1 mM. With these simulation parameters, the steady-state mean

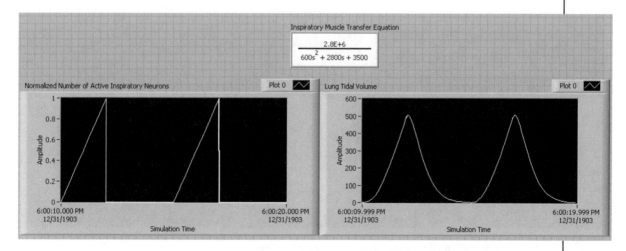

FIGURE 17.9: Front panel for inspiratory muscle simulation.

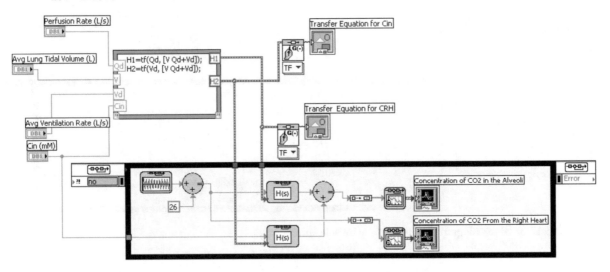

FIGURE 17.10: Block diagram for Milhorn lung equation simulation.

for the alveolar CO– concentration was roughly 16.625 mM (Figure 17.11). The results correspond with Sun et al.'s [6] experimental results.

Lacking reliable values for the CO_2 generation rate in the tissue and body compartments or any published values for comparison, a LabVIEW simulation for the tissue model was generated with user-supplied parameters for future use. The block diagram for this simulation is shown in Figure 17.12.

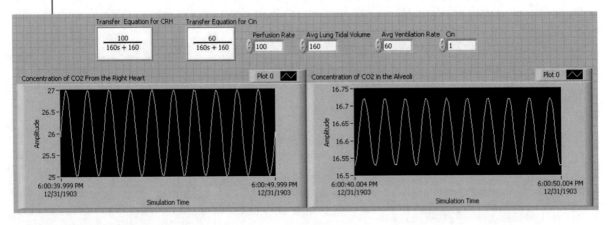

FIGURE 17.11: Front panel for Milhorn lung equation simulation.

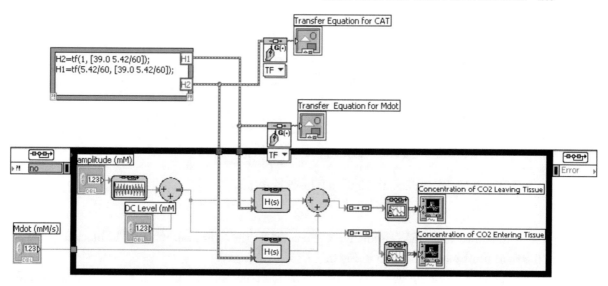

FIGURE 17.12: Block diagram for simulation of tissue model.

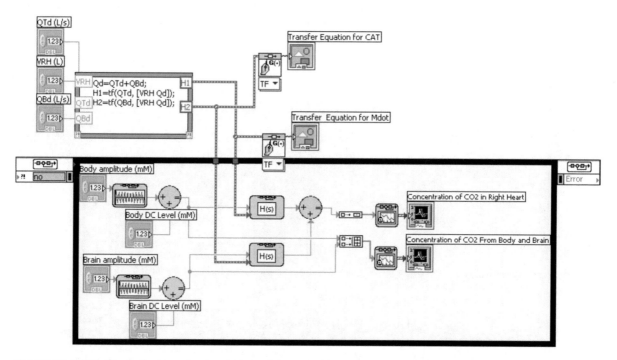

FIGURE 17.13: Block diagram for simulation of the right heart model.

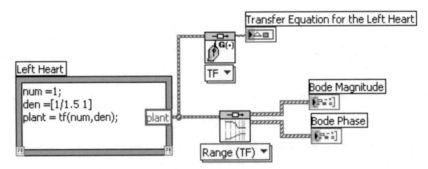

FIGURE 17.14: Block diagram for the left heart Bode plot.

A similar LabVIEW simulation was created for the right heart model. The block diagram for the right heart model is shown in Figure 17.13.

Bode plots for the right heart and the chemoreceptors were also generated using LabVIEW. The Bode plots for the left heart model were generated using the block diagram in Figure 17.14; the LabVIEW results are shown in Figure 17.15. Although the left heart model included a time shift in the numerator, the time shift was not replicated in LabVIEW program.

The Bode plots for the central and peripheral chemoreceptors were generated using the block diagram shown in Figure 17.16; the resulting LabVIEW front panel output, Bode plots, are shown in Figure 17.17.

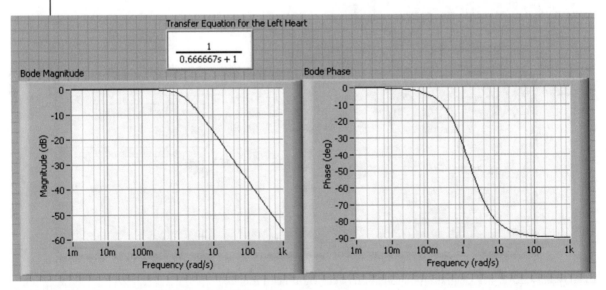

FIGURE 17.15: Front panel for left heart Bode plot.

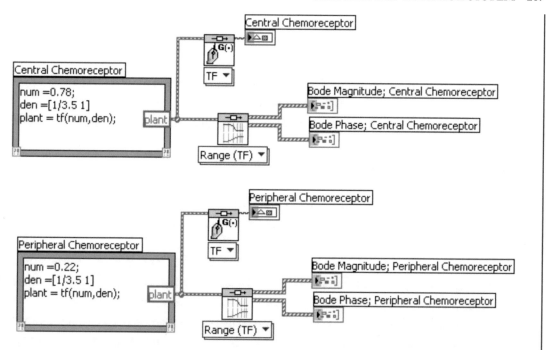

FIGURE 17.16: Block diagram for Bode plots of central and peripheral chemoreceptors.

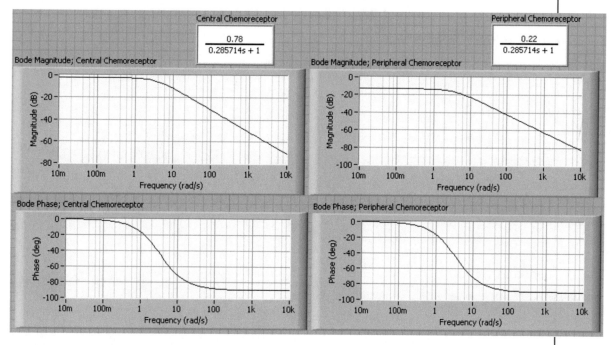

FIGURE 17.17: Front panel for bode plots of central and peripheral chemoreceptors.

REFERENCES

[1]   Milhorn, H. T., Jr., *The Application of Control Theory to Physiological Systems*, W.B. Saunders Company, Philadelphia (1966).

[2]   Po, H. N., and Senozan, N. M., Henderson–Hasselbalch equation: its history and limitations, *J. Chem. Educ.*, **78**, 1499–1503 (2001).

[3]   de Levie, Robert, The Henderson–Hasselbalch equation: its history and limitations, *J. Chem. Educ.*, **80**, 146 (2003).

[4]   Silverthorn, D. U., *Human Physiology*, 4th ed., Pearson Education, Inc., San Francisco (2007).

[5]   Topor, Z. L., Vasilakos, K., Younes, M., and Remmers, J. E., Model based analysis of sleep disordered breathing in congestive heart failure, *Respir. Physiol. Neurobiol.*, **155.1**, 82–92 (2007).

[6]   Sun, X. G., Hansen, J. E., Stringer, W. W., Ting, H., and Wasserman, K., Carbon dioxide pressure concentration relationship in arterial and mixed venous blood during exercise, *J. Appl. Physiol.*, **90.5**, 1798–1810 (2001).

[7]   Ursino, M., and Magosso, E., A theoretical analysis of the carotid body chemoreceptors response to O_2 and CO_2 pressure changes, *Respir. Physiol. Neurobiol.*, **130.1**, 99–110 (2002).

Anthony Salvaggio and Brian Schepp are acknowledged for their contributions toward the development of this chapter.

· · · ·

Author Biography

Charles Lessard, Ph.D., Lt. Colonel, U.S. Air Force (Retired), IEEE Life Fellow, Associate Professor, Biomedical Engineering Department, Texas A&M Univesity. Dr. Lessard received his doctorate in Electrical Engineering from Marquette University in 1972, his master's degree in Electrical Engineering from the Air Force Institute of Technology in 1967 and his bachelor's also in Electrical Engineering from Texas A&M University in 1958. As a U.S. Air Force Pilot, he flew F-86L All Weather Interceptors and B-52 Strategic Bombers. He was a Lead Engineer and scientist with the School of Aerospace Medicine and the AF Medical Research Laboratories before serving as the overseas Program Manager and Spain's Field Office director in the installation and testing of Spain's Air Defense system. He joined the Texas A&M Engineering faculty in 1981 as an Assistant Professor after retiring from the Air Force in July 1981. Dr. Lessard has served as an Adjunct Professor with the Air Force Institute of Technology, Electrical Engineering, Assistant Professor, Night Division, San Antonio College, Assistant Professor, Night Division, St. Philips College and Adjunct Professor, Physical Therapy, Texas Women's University. Dr. Lessard specializes in physiological signal processing, design of virtual medical instrumentation, control systems, noninvasive physiological measurements, vital signs, nystagmus, sleep and performance decrement, spatial disorientation, acceleration (G)-induced loss of consciousness (G-LOC), and neural network analysis.

Printed in the United States
by Baker & Taylor Publisher Services